HOW WILL WE MEET THE ENERGY CRISIS?
Power for Tomorrow's World

Today our growing need for power is fast outstripping our means of producing it—but further expansion of our current energy sources poses a dangerous threat to the environment. A fascinating examination of what is being done, and the wonders the future may hold as scientists work to turn their visions into realities.

TOMORROW'S WORLD SERIES

CLEAN AIR—CLEAN WATER FOR TOMORROW'S WORLD

HOW WILL WE FEED THE HUNGRY BILLIONS?
Food for Tomorrow's World

HOW WILL WE MEET THE ENERGY CRISIS?
Power for Tomorrow's World

HOW WILL WE MOVE ALL THE PEOPLE?
Transportation for Tomorrow's World

HOW WILL WE MEET THE ENERGY CRISIS?

Power for Tomorrow's World

by Reed Millard
and the editors of
Science Book Associates
with photographs

JULIAN MESSNER NEW YORK

Published by Julian Messner
a division of Simon & Schuster, Inc.
1 West 39th Street, New York, N.Y. 10018
All Rights Reserved

Copyright, ©, 1971 by Simon & Schuster, Inc.

Printed in the United States of America
ISBN 0-671-32477-2 Cloth Trade
0-671-32478-0 MCE
Library of Congress Catalog Card No. 78-161514

WE WISH TO THANK the many organizations in government and private industry whose assistance has made this book possible. We are particularly grateful to many persons associated with the following:

American Electric Power Service Corp., American Gas Association, American Petroleum Institute, Argonne National Laboratory, Arthur D. Little, Inc., Atomic Industrial Forum, Avco Everett Research Laboratory, Bell Laboratories, Canadian Consulate General, Churchill Falls Corporation, Ltd., Combustion Engineering Company, Edison Electric Institute, El Paso Natural Gas Company, French Embassy Press and Information Division, General Electric, Independent Natural Gas Company, Lawrence Radiation Laboratory, Los Alamos Scientific Laboratory, Massachusetts Department of Commerce, National Aeronautics and Space Administration, National Coal Association, National Rural Electric Cooperative, New York State Electric and Gas Corp., Pacific Gas & Electric Company, Ralph M. Parsons Company, Power Reactor Development Company, Raytheon Company, Sandia Laboratories, Solar Energy Society, Southern California Edison Company, the United Nations, U.S. Army Natick

Laboratories, U.S. Atomic Energy Commission, U.S. Bureau of Mines, U.S. Bureau of Reclamation, Westinghouse Electric Corporation.

THE EDITORS,
SCIENCE BOOK ASSOCIATES

CONTENTS

1	The Energy Crisis	9
2	Fuels for Power Without Pollution	16
3	Power From the Atom	36
4	How Safe Are Atomic Power Plants?	55
5	New Ways to Deliver Power	68
6	Miniature Power Plants	82
7	New Power From Rivers	105
8	Power From the Sea	125
9	Power From the Earth	145
10	Power From the Sun	163
	Epilogue	180
	Suggested Further Readings	183
	Sources of Information About Power	185
	Index	187

1
THE ENERGY CRISIS

Blackout!

A great city comes to a sudden halt. It could be New York, Chicago, Los Angeles or London. Elevators stall, appliances whine to a stop and computers lose their memories. Above the airports, the pilots of arriving jet liners see the guiding lights on the runways disappear. Terrifying and destructive, blackouts shock the people who experience them.

Fortunately, they do not happen often. Blackouts, which occur because of some breakdown in the power delivery system, and brownouts and dimouts, which result from a power shortage, force us to think about our dependence on electrical power. It would be hard indeed to imagine our world without it.

How Will We Meet the Energy Crisis?

Yet, as power shortages in the late '60s and early '70s demonstrated, *we could run out of electricity.*

Look at the background for a situation which few people would have thought possible a few years ago. The amount of electricity used in the United States has doubled every decade since the beginning of the century. Today the United States produces more electricity than the next four countries combined—that is, more than that produced by the Soviet Union, Japan, the United Kingdom and West Germany. We have harnessed rivers with giant power-producing dams; we have used up billions of tons of coal, millions of gallons of oil, huge quantities of natural gas and, recently, trainloads of uranium ore for nuclear power plants.

Yet we've just begun. In the next ten years, by the most conservative estimates, we will need as much electricity as we have used in all the years since Thomas Edison invented the incandescent light bulb, almost a hundred years ago!

And this prodigious increase is just that caused by the normal, predictable climb in electricity usage resulting from a bigger population, increased use of appliances and greater industrial requirements. It is not even taking into account the wholly *new* demands for electrical power—huge new electrified mass-transit systems, desalination plants that will turn salt water into fresh and the raising of food for

The Energy Crisis

our hungry world in indoor "farms" requiring artificial light, air conditioning and heating.

Many planners believe we'll go beyond even these electricity-demanding activities and build totally electric cities. Dr. Athelstan Spilhaus, President of the Franklin Institute, visualizes such a community, proposing that one be built as an "Experimental City." His plans for it represent a glowing vision of desirable living. As the veritable beating heart of the city he describes a power plant—most likely a nuclear power plant—capable of providing the enormous amount of energy required to realize such a dream.

Spilhaus sees it as a city without slums, traffic jams, noise, smog, water shortages, pollution, industrial blight, junk yards, suburban sprawl or urban jungle—without unemployment and without the social and economic conflicts that cause strife. It would be a city whose citizens do not have to pay dearly for the barest comforts of life.

"It will be centrally heated and possibly air conditioned," he explains. "It will require no external water supply, since all of its water, for whatever purpose, will be recycled and reused. No fuel-burning automobiles or plants will pollute the atmosphere. Public and private vehicles will be electrically driven."

In the Spilhaus Experimental City, the need for

How Will We Meet the Energy Crisis?

electrical power will be vastly greater than in any city of today. The central heating and air conditioning will be electrically operated; the water and sewage recycling systems will use electricity; recharging the batteries of electric cars will take still more.

Whether such dream cities are actually built or not, great new demands for electricity can certainly be predicted—demands that will create an energy crisis.

There are two basic problems. One is the looming shortages of fuels to run power plants. While nuclear power is making rapid advances and while, as we'll see in this book, there are other promising sources of power which we may be able to tap, most power

Can we produce enough extra power to meet the greatly increased demands of tomorrow's electrified cities?

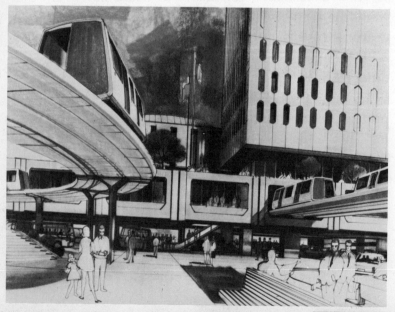

WESTINGHOUSE

today comes from fossil fuels. Coal, oil and natural gas provide 80 per cent of all power produced in the United States, and for a good many years to come they will still be the major providers of electricity. Yet, in 1969 and 1970, the amount of natural gas we used was greater than the amount located in new discoveries. In other words, we had to dip into known reserves. Coal wasn't dug out of the ground fast enough or delivered fast enough, so many power plants found themselves with only a few days' supply, instead of the two to three months' stockpile considered necessary. In the winter of 1970-71, many of the power plants which use oil could not get enough to keep operating at full production.

Shortages of fuel may be troublesome, but there is an even greater cause for worry. Our need for electricity is on a collision course with the need to protect our air and water from pollution. The unfortunate fact is that the most commonly used power plant fuels—coal and oil—spew vast quantities of dangerous, corrosive chemicals into the air. Although automobiles with internal combustion engines are worse offenders, power plants produce many of the most virulent pollutant substances, adding them to the choking smogs that threaten our cities. And, while they do not poison our waters with the chemicals dumped in them by some factories, they do damage the water with a different kind of pollutant—heat. Even atomic power plants, which do not pollute the air, are destructive to the water near them. Moreover, many environmentalists are worried about dangers the power plants may present when built in congested urban areas. Some scientists are haunted by the fear that an accident could have deadly results.

Thus the task of producing enough power for our growing electrical needs is not a simple one of finding, extracting and transporting fuel. It is also a problem of finding ways to keep the fuels from hurting the environment. In 1970, S. David Freeman, of the White House Office of Science and Technology, a government organization which advises the president on pressing problems, said: "Environmental problems may be the straw that broke the camel's back."

No wonder scientists for government, industry and international organizations—including the United Nations—are joining in a world-wide drive to find solutions to the mounting power crisis. It is important to recognize that this crisis concerns not only the United States and the highly developed countries which now use the most electricity, but the developing nations which, as they raise their standard of living, will also need more and more electricity. They, in turn, will place demands on the world supply of fuels, demonstrating again that in the world today the big problems are not local, national or even continental; they are world-wide.

The Energy Crisis

Can science and technology meet the energy crisis? Can we get all the electricity we will need in ways that will not damage our environment? No one can answer these questions for sure, but as we survey the new tools that can produce power, you may decide that there is reason to think that the planners are not wildly wrong in their hopes for a brighter, electrified world of tomorrow.

2
FUELS FOR POWER WITHOUT POLLUTION

Flames race through a coal mine, deep underground. No one makes any attempt to fight the raging fire. In fact, the men on the surface actually encourage the underground conflagration by feeding oxygen to speed up the burning. Strange? No, because this is no disaster. It is an experiment, in which scientists have deliberately set fire to a coal mine.

What does this event have to do with the most pressing problem facing power producers looking for ways to produce power without pollution? It has a lot to do with it, because the success of this experiment in an Alabama coal mine can help give us breathable air *and* the power we need. It's one step

Fuels for Power Without Pollution

along the way toward making coal a friend, not an enemy, of the environment.

Today, fossil fuels—coal, oil and gas—provide the heat that makes the steam that turns the turbines that create more than three-fourths of all electricity produced in the early 1970s. This has been true for decades. It is likely to remain true for decades to come, in spite of the advances in atomic power and other sources of power we will be looking at later in this book.

Coal—the Big Power Producer

Coal, the marvelous black stone that burns, provides almost 60 per cent of all the electrical power produced in the United States. How much of this fuel do we have left? Geologists estimate world coal reserves at *5,000,000,000,000 tons*. About 40 per cent of this immense store of energy is located in coal-rich North America. At present the United States is using only a small fraction of its deposits—less than 600,000,000 tons a year. Some estimates say we could use coal at the present rate for 2,000 years without exhausting the supply.

The big problem, therefore, is not one of quantity. The trouble with coal has always been that burning it pollutes the air. The pollutants come in three forms—particulate matter (dust), sulfur oxides and nitrogen oxides. Each of them presents its own

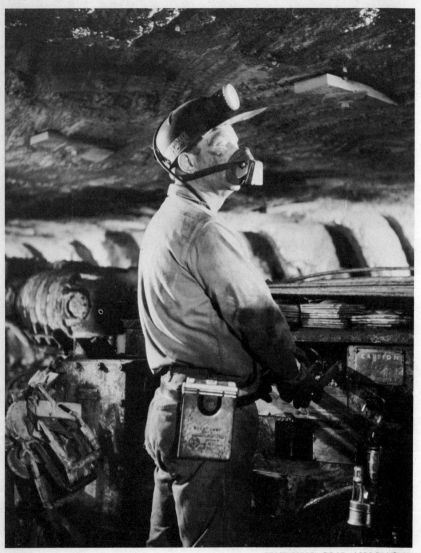

NATIONAL COAL ASSOCIATION

Coal—the "black sunlight"—provides more than half the fuel that creates today's electricity. It promises to be an equally important fuel for tomorrow's power.

Fuels for Power Without Pollution

dangers to the environment—to plants, animals and human beings subjected to them.

Particulates are tiny particles—as small as 1/12,000 inch—which can reach deep into the lungs. Here they act as carriers for dangerous chemicals which otherwise could not get past the mucous which absorbs gaseous chemicals.

Sulfur oxides are heavy, pungent, colorless gases. A major one, sulfur dioxide, damages bodily tissues and is suspected of being a cause of cancer.

Nitrogen oxides are compounds formed in a union between nitrogen and oxygen. They can take a number of forms, including nitrogen dioxide, one pollutant which you can see in the atmosphere. It helps give urban smog its yellow-brown color.

The fact that burning coal emits these dangerous substances doesn't mean we have to stop using it as our main producer of electricity. Scientists and engineers have found many promising ways to take the pollution out of this energy-laden fuel.

One approach is to try to extract the pollutants from the coal as it burns in the power plant furnaces or rushes up the smokestacks. Filter systems have been developed that are capable of capturing 99 per cent of the particulates, but, unfortunately, the researchers have not had the same kind of success in removing the sulfur and nitrogen compounds. An experimental plant in Everett, Massachusetts, being constructed in 1971, may prove that it can be done, but in 1970 the Academy of Engineers of the National Research Council reported that "a commercially proven technology for the control of sulfur does not exist."

The brightest future for getting pollutants out of coal at the power plant probably lies in a new way of turning fuel into electricity—MHD. (The letters stand for *m*agneto*h*ydro*d*ynamics.) This process utilizes hot gases obtained from the burning of coal or oil at extremely high temperatures. These temperatures are so great that most of the pollutants in the fuel are burned away. You'll be hearing a lot about MHD in years to come, for it promises not only pollution-free power but also conservation of natural resources.

The story of MHD began more than 150 years ago in the laboratory of the tireless electrical pioneer Michael Faraday. In early experiments in generating electricity, he learned that hot gases served the same purpose as copper coils. Electricity could be produced by moving hot gases through a magnetic field. After generators of the type in use today in all power plants were developed, scientists were always aware of this discovery. They knew that if flowing hot gases could be substituted for the rotating copper coils of conventional generators, this would be a far more efficient means of generating electricity. However, it remained only a tantalizing theoretical possibility because of a seemingly insuperable problem.

NEW YORK STATE ELECTRIC & GAS CORP.

This giant coal-burning power plant is located near mines in Pennsylvania. It is easier to carry the electricity by high tension wire than it is to transport the coal.

How Will We Meet the Energy Crisis?

To work properly, the gases had to be extremely hot —around 5,000° F. If they were cooler, not enough electricity would be developed.

Only recently have advances in plasma physics (a gas at extremely high temperatures is called a "plasma"—the fourth state of matter) made it practical to deal with such high temperatures in a power system. Today, in the Soviet Union, West Germany, Japan and other countries, as well as in the United States, great strides are being made with experimental MHD systems. In the United States, the MHD pioneering firm of Avco, with the Edison Electric Institute and several New England utilities, has constructed a test plant. The performance of this and other experimental facilities demonstrates the possibilities of MHD as the power of the future. Its efficiency ran as high as 60 per cent, compared with 40 per cent for ordinary power plants. This means that MHD can produce 4,500 kilowatt-hours of electricity from one ton of coal, compared with 3,000 kilowatt-hours by conventional methods.

The most impressive performance of MHD is its ability to fight air pollution. On the basis of experiments, it is calculated that a conventional coal-burning steam power plant of 1,000,000-kilowatt capacity emits 33 tons of particulate matter a day; a coal-burning MHD plant emits only three tons. From the stacks of a steam plant pour 80 tons of nitrogen oxides—from an MHD plant only four tons. The

Fuels for Power Without Pollution

most spectacular improvement is in the quantity of sulfur oxides: only three tons from an MHD plant instead of the huge quantity from a steam plant—450 tons! No wonder many environmentalists and power experts believe that MHD holds the secret for cleaning up coal.

In the meantime, other expedients are being tried. One obvious approach is to locate coal-burning power plants far from population centers. Since the largest coal deposits in the United States are in the wide reaches of Western states, this seems, offhand, to be a desirable scheme. You just build the power plant right by the mine in a remote area, and save all the expense of transporting the coal a long distance.

The trouble is, this simple line of reasoning hasn't been worked out as a successful antipollution measure. A good illustration is the biggest "mine-mouth burning" project in the United States, located in the remote Four Corners country, where Colorado, New Mexico, Arizona and Utah meet. Headed by Southern California Edison, a group of power companies called WEST (Western Energy Supply and Transmission Associates) has built huge power plants on the Navaho Indian Reservation in New Mexico. Here two 750,000-kilowatt steam turbine plants sit right above the coalbeds. Power from the plants is carried by high-tension lines across the deserts and mountains to southern California.

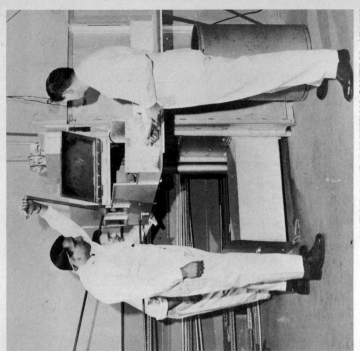

NATIONAL AIR POLLUTION CONTROL ADMIN.

Technologists work on the problem of reducing pollution from power plants. This experimental device tests different methods of combustion.

KOPPERS COMPANY

Electrostatic devices such as these being tested before installation can remove many of the pollutants that would otherwise go up the stacks of coal-burning power plants.

Fuels for Power Without Pollution

Unfortunately, pollutants from the plants are also carried long distances. In Albuquerque, 165 miles away, fine dust from their high stacks has raised pollution levels. In many parts of northern Arizona, the plants have been charged with creating high clouds which have reduced the sunlight of that sunny area by 15 per cent. Environmental groups have entered strong protests against the building of more giant power plants at this site. Similar experiences with mine-mouth operations in Pennsylvania and West Virginia indicate that distance is no cure for power plant pollution.

That's where that deliberately set fire in a coal mine comes into the picture. It's one way to turn coal into a clean gas. There's nothing very new about the gasification of coal. Coal was the source of the gas used in homes during the 19th century, and well into the 20th. In the late 1930s welded steel pipes of large diameter made it possible to carry natural gas long distances. The once familiar "gas house," where coal was turned into gas, disappeared from the American landscape.

Today's technology has come up with some new ways of getting the gas out of coal and piping off the hot gases. The U.S. Bureau of Mines and the Office of Coal Research are conducting pilot plant operations. One plant in Chicago uses tall reactor towers, making it look something like an oil refinery. It extracts many gases from the coal, among them hydrogen and carbon monoxide, which are used at the plant to power a generator that makes some of the electricity used to operate the plant.

The most exciting prospect for the future is the bold approach of simply gasifying the coal in the mine. Many technical difficulties lie in the way of adopting this scheme, but in full-scale tests in Alabama mines, engineers have proved that it works. The coal in a mine is fired, oxygen is piped down to it and the resultant heat makes coal buried deep under the earth give up its valuable gases, which are then piped off. The mine, in effect, becomes an enormous retort for cooking the coal. With the expense of mining the coal eliminated, experts estimate that gas may be produced for less than half the cost of digging up the coal and processing it above ground.

What About Oil?

When used in power plants, oil presents much the same problem as coal. Most petroleum contains sulfur compounds. These can be removed by chemical processes at refineries, but the expense of removal adds to the cost of the oil. As in the case of coal, they can also be removed at the power plant by devices that process fumes before they escape from the smokestacks.

No matter how successful technologists may be in

U.S. BUREAU OF MINES

Changing coal to gas is a complicated process being carried out in this experimental plant. In the future this process may be accomplished by burning coal right in the mine.

getting the sulfur out of oil, it is doubtful if this versatile fuel can contribute much to the needs of power plants. We need so much of it to fuel automobiles, heat homes and serve as raw material for petrochemicals that there just isn't likely to be enough of it to go around. To be sure, a world-wide search for oil is turning up new supplies, such as the vast new oil field on the North Slope of Alaska. Offshore drillers are at work on the continental shelves of our own continent and others. Big finds have been made off California, Alaska and Australia and near several other coasts, with the promise of many more underwater oil fields yet to be tapped.

Perhaps the greatest treasure trove will be found in deposits that have not yet been exploited—those locked in sand or rock. Some of this petroleum is trapped in "oil sand." Along the Athabasca River in western Canada, there is a giant bed of this oil-saturated sand. Geologists estimate that this deposit may cover 30,000 square miles and contain 250,000,000,000 barrels of oil.

The greatest oil reserve of all is found in "oil shale," a sedimentary rock that has oil locked inside it. When heated in a furnace to some 700° F., oil shale gives up its oil. Nuclear scientists envision a bold scheme to extract oil from shale. As part of Project Plowshare, a program designed to explore the use of nuclear explosives for peacetime purposes, they propose to explode a hydrogen "bomb"

ARABIAN AMERICAN OIL COMPANY

Oil can be used as a fuel in steam power plants, but most of it is needed for automobiles, home heating and the making of petrochemicals. Even new discoveries of offshore oil may not produce enough to go around.

An atmospheric scientist measures the amount of pollutants distributed in the atmosphere by nearby power plants and factories.

DU PONT

deep inside a shale deposit. The mighty force and heat will, they believe, cause the oil in the shale to liquefy so that it can be pumped to the surface.

Natural Gas—Pollution-Free Fuel

Natural gas, which has been termed the "magic flame that works for man," has long been a valuable fuel for producing heat. The ancient Greeks thought that natural gas escaping from rocks was the breath of the mighty god Apollo. The Chinese, more than 3,000 years ago, actually put it to practical use by channeling this "burning spirit" from a natural gas well through bamboo pipes to brine wells where they used it for evaporating salt.

In the Western World, gas from coal was used long before anyone found a way to utilize the deposits of natural gas that exist in the earth. In the United States, George Washington was the owner of the first discovered natural gas well. Located in what is presently Kanawha County, West Virginia, its fumes were a mere curiosity, and neither George Washington nor anyone else thought of making any practical use of them. Later, gas wells were discovered near Lake Erie, and one at Fredonia, New York, was the first to be actually used. It was harnessed to light the street lamps of that city in 1823.

Since then natural gas has found an astonishing range of uses in modern industrial society. It serves

Fuels for Power Without Pollution

as a raw material for more than 2,500 different products, ranging from nylon to fertilizers. It heats 40,000,000 homes. And it's the fuel used in some power plants—but not in nearly enough to satisfy the environmentalists, who would like to see this pollution-free fuel used in a lot more.

The trouble is that we're burning it up at the rate of more than 9,000,000,000,000 cubic feet a year. And, with only 200,000,000,000,000 cubic feet of proved reserves that can be extracted by present methods left in the United States, there is reason to fear that we could run out of natural gas. However, thanks to atomic energy, we may be able to tap immense deposits of natural gas not included in the proved reserves.

Locked in what geologists call "tight" formations of rock, there is an estimated 317,000,000,000,000 cubic feet of natural gas. The problem is to get it out. Unlike the gas in underground reservoirs, which can readily be reached by drilled wells, this gas is trapped in the pores of rocks. Long ago engineers looked for ways of making the rocks give up this gas. In a process called stimulation, they tried setting off explosives in gas wells. This created fractures in the rocks and freed some of the trapped gas. Later it was found that forcing down liquids under high pressure had the same effect. Unfortunately, the results were limited. The stubborn rocks still retained most of the gas.

How Will We Meet the Energy Crisis?

In December, 1967, the El Paso Natural Gas Company and the Atomic Energy Commission carried out a pioneering Plowshare experiment. For the first time in history, the mighty power of nuclear explosives was actually put to work for a peacetime purpose. For Project Gasbuggy, the engineers drilled a hole 4,200 feet deep into a layer of sandstone near Farmington, New Mexico. They lowered a 26-kiloton package of nuclear might to the bottom of the hole. The 15-foot-long steel cylinder was hardly impressive, but it contained more explosive force than the Hiroshima bomb.

When the device was detonated, observers on the ground saw and felt little. But deep within the earth, the awesome fury of nuclear explosives, in a millionth of a second, vaporized, melted and crushed the surrounding rock. The explosion formed a spherical cavity that ballooned outwards to a diameter of more than 150 feet.

The force of the explosion also fractured rock beyond the cavity for approximately 650 feet on all sides. Then the ceiling of the cavity collapsed, enlarging the cavern to a huge cylindrical chimney 350 to 400 feet high.

What was all this for? The scientists figured that natural gas would flow from the fractured rock into this enormous chamber, forming an underground "tank" of natural gas that could easily be pumped out later. Their calculations proved to be correct.

Fuels for Power Without Pollution

Gasbuggy has shown that the natural gas yield from "tight" formations can be increased by as much as 700 per cent.

A later experiment, Project Rulison, carried out late in 1969, was even more successful. Some 8,000 feet down under the Colorado plateau, scientists used nuclear power to attack a particularly stubborn gas field. An enormous amount of gas was locked in the rock formations of this area. After the 40-kiloton blast was set off, it "shook loose" so much gas that when the well was tapped, in late 1970, the yield was as high as 20,000,000 cubic feet a day—compared with a yield of 100,000 cubic feet from nearby wells. Of course, such a yield did not continue, but a safe estimate puts the amount of gas released soon after the blast at 1,000 per cent of the yield that could otherwise be expected.

No one can be sure that the use of atomic explosives is the magic key that will force the earth to yield enough natural gas to use in power plants, as well as in the many other tasks it performs. Many problems remain to be solved, many questions to be answered. Will it cost too much to get gas this way? Can we be sure that, if nuclear blasting is used on a large scale, radioactivity won't contaminate the soil and water around the blasts or the air above them? Will the gas itself be as free from radioactivity as the Gasbuggy and Rulison experiments seem to indicate?

LOS ALAMOS SCIENTIFIC LABORATORY

Project Rulison technicians are lowering a nuclear explosive into a hole drilled in a gas field in Colorado. The explosion, 8,442 feet below the ground, will increase the flow of natural gas that would otherwise be trapped in rocks.

NORTHERN ILLINOIS GAS CO.

Natural gas carried through a network of pipelines provides clean-burning fuel for millions of homes. Will we be able to find enough of this versatile fuel to use in power plants?

Many scientists are doubtful. They fear that the answers to these and many other questions about nuclear blasting will not be encouraging. Some experts believe that further experiments should be called off. Others, like Dr. Edward Teller, staunchly maintain that our need for energy is so desperate that we must push forward with any promising efforts to tap new energy sources.

In any case, it would be hard to find any expert who would not agree that the ideal way to get energy for power production is from the clean-burning flame of natural gas.

3
POWER FROM THE ATOM

The months-long Arctic night has settled down, and the sun has dropped below the horizon. But glowing in the vast reach of ice and snow is a brilliant area of light—a domed city.

Its streets are lit with the full brightness of summer sunlight. Flowers bloom in the parks, birds sing in the trees, people stroll the walks in shirtsleeves. Children play on a beach of white sand beside a small lake.

In brightly lit factories workers process the minerals brought up from the rich mines under this city in the far north.

What makes it possible?

Electricity in unlimited quantities—electricity wrested from the nucleus of the atom.

Power From the Atom

While you may never live in, or even visit, a domed city in the Arctic, you almost certainly will live in a city much more electrified than today's. And it is very likely, too, that most cities of tomorrow will get their power from nuclear sources. To be sure, for some time to come, fossil fuels—coal, oil and natural gas—will continue to supply much of our power needs but most experts agree that we will come to depend more and more on atomic energy. The only long-range rival for the mighty power locked in the atom is the almost limitless energy of the sun itself.

Nuclear Reactors for Today and Tomorrow

You may think of the atom bomb as the first use of atomic energy. Actually, the pioneers of nuclear science built a nuclear power plant before they succeeded in making the first atomic bomb.

It all began in what had once been a squash court beneath the stadium at the University of Chicago. Here a band of 42 scientists gathered on a cold December day in 1942. Among them were Lise Meitner, who was the first to actually split an atom, and Enrico Fermi, who had won the Nobel prize in 1938 for his pioneer work in nuclear physics.

As they gathered on a balcony above what looked like a giant beehive made of black bricks, the scientists were tense. What happened inside that pile of graphite bricks would determine whether sci-

POWER REACTOR DEVELOPMENT CORP.

The many dials and gauges in the control room of a nuclear power plant monitor the complex workings of the reactor and generating units.

ence really could produce usable atomic energy.

What the theory behind this effort added up to was this: If enough uranium were put in one place, the neutrons naturally released by fissioning uranium atoms would stimulate other uranium atoms to fission, releasing more neutrons, causing more atoms to fission and so on. If the rate of energy release were rapid enough, the atom could be a new power source, providing energy on a scale never before dreamed of. The scientists knew that such energy could be used in bombs of terrible destructive force. They also knew that if it could be controlled, and released at a slower rate, it could be a new source of energy for peaceful uses.

Power From the Atom

To test the theory, many tons of uranium had been mined, processed and brought to Chicago. Gathering this quantity of uranium had been a difficult scientific project in itself, involving the work of many men and companies.

To keep the chain reaction from starting prematurely, cadmium rods had been inserted into the heart of what the scientists called "the pile," since this reactor was actually a pile of bricks. The word "pile" stuck and became the term applied to atomic reactors, even though they look nothing like this one. The cadmium rods "absorbed" the neutrons given off by the splitting uranium atoms and prevented them from causing other atoms to split. When the cadmium rods were withdrawn from the pile, the chain reaction would begin.

It was 9:45 A.M. when Fermi ordered the first rods withdrawn. These were operated by a small electric motor and were designed as safety rods, to be dropped back into place if the reaction got out of hand. Next, the main emergency rod was withdrawn. It was tied to a rope which was fastened to the balcony on which the scientists stood. One held an ax, ready to cut the rope and let the rod fall back into the pile. This safety measure was taken in case the electrically operated rods failed.

At this point only one control rod kept the chain reaction from beginning. Inch by inch, foot by foot, it was removed, as the scientists watched the dials of counters which measured radiation escaping from the pile. Each step of the way, the counters clicked faster. Fermi and the other scientists performed rapid computations on their slide rules. The experiment was working as they had figured.

All through the morning, the slow withdrawal of the rod continued. The clicking of the radiation counters, growing more rapid, was the loudest sound in the room.

After a lunch break, the scientists came back and continued the withdrawal of the rod. At 3:45 in the afternoon, Fermi said: "Pull it out another foot. This is going to do it."

The clicking of the radiation counters was so fast that it was a steady hum of sound. If it did not level off, it would mean that the chain reaction had begun. After another minute, the count was still rising.

"The reaction is self-sustaining," Fermi said. His voice betrayed none of the excitement he must have felt, but there was a broad smile on his face.

The scientists watched in silence, staring at the motionless bulk of the world's first atomic reactor. Inside it, uranium atoms were splitting into pieces, releasing neutrons.

It was an awesome moment. As Fremi said later, "It meant that the release of atomic energy on a large scale would be only a matter of time."

Thus, from the very beginning, though they had to turn aside from the momentous—and frightening—task of developing the atom bomb, the scientists of the Manhattan Project dreamed of tapping the mighty power of the atom for peaceful purposes.

From Atoms to Electricity

There is nothing mysterious about the way atomic energy is used to produce electricity in the hundreds of nuclear power plants already in operation—or in the thousands that will be helping to electrify tomorrow's world. A nuclear power plant is similar to a conventional thermal power plant. Each type uses steam to drive a turbine generator that produces electricity. The heat energy of the steam is converted to mechanical energy in the turbine, and the generator then converts the mechanical energy into electrical energy, or electricity. Although the turbine functions equally well no matter where the steam comes from, the origin of the steam is important to us, for it is here that nuclear and conventional plants differ.

How is steam produced? Conventional plants burn coal, oil or gas, and heat from the combustion of these fossil fuels boils water to make steam. In nuclear plants, on the other hand, no burning or combustion takes place. Nuclear fission is used in-

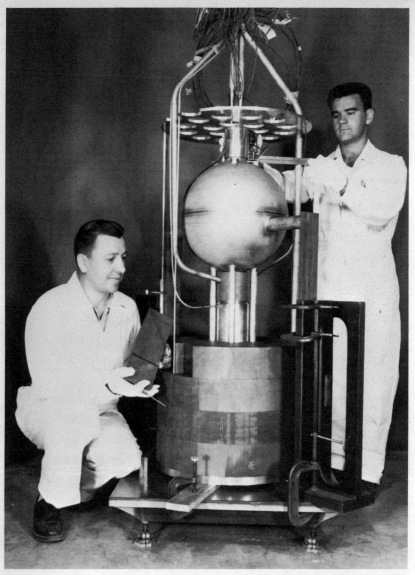

ATOMICS INTERNATIONAL

Whether they are small, like this research reactor. . . .

UNITED KINGDOM ATOMIC ENERGY AUTHORITY

.... or giants like the one housed in this huge power plant ....

.... The workings of reactors are basically the same.

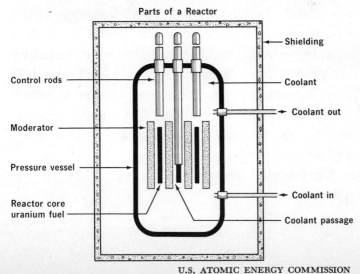

U.S. ATOMIC ENERGY COMMISSION

stead. The fission reaction generates heat, and this heat is transferred, sometimes indirectly, to the water that produces the steam.

The fission process requires a particular kind of heavy element, such as uranium or plutonium, as a basic material. Let us consider uranium. Natural uranium is a mixture of three isotopes, atomic forms which are chemically alike but which vary in mass. An atom of one of these isotopes, uranium^{235}, can readily undergo fission when a free neutron (an energetic subatomic particle) strikes its heavy central nucleus. The nucleus breaks into two pieces, which fly apart at high speed; in addition, two or three new neutrons are released.

In the reactor the kinetic energy of the flying fission fragments is converted to heat when they collide with surrounding atoms, and the released neutrons cause a chain reaction by initiating new fissions in other atoms of the nuclear fuel.

More than 30,000,000,000 fissions must occur in one second to release each watt of energy. If the chain reaction is to be useful, the fissions must occur at a desired rate, and the heat that is generated by the process must be removed. The job of a nuclear reactor, then, is to provide an environment in which fission reactions can be initiated, sustained and controlled and to make possible recovery of the resultant heat.

The essential components of a reactor are:

—the fuel, which fissions to produce neutrons and to release energy;
—the control elements, which are used to set the energy release rate; and
—the cooling fluid, which removes the heat generated in the reactor.

Engineers have worked out a number of different ways to use the heat produced by fissioning atoms. For example, this is the way it worked in the world's first commercial nuclear power plant. This pioneering reactor, which began operation at Shippingport, Pennsylvania, in 1957, was not vastly different from the first crude pile of graphite bricks built by Fermi and his associates 15 years before. But it embodied many improvements in design that allowed a much smaller quantity of uranium fuel to achieve a chain reaction.

The Shippingport reactor is what nuclear engineers call a pressurized-water reactor. In this kind of reactor, water under pressure is circulated through the heart of the reactor, where the chain reaction is taking place. The fissioning atoms generate great heat, raising the water to temperatures of many hundreds of degrees Fahrenheit. (The pressure keeps the water from boiling, just as the pressure in a home pressure cooker prevents water inside from boiling, even though its temperature rises far above 212° F.—the "normal," sea-level boiling point of water.)

How Will We Meet the Energy Crisis?

After being heated in the nuclear furnace, the water circulates to a heat exchanger. There it runs through a network of tubes, around which flows unpressurized water. Heat transfers from the pressurized water to the unpressurized water through the walls of the tubes.

Variations of this system are actually at work in the nuclear power plants being built in all parts of the world. In them every gram of fissioned uranium produces about 7,000 kilowatt-hours of electricity. This tiny fraction of an ounce produces as much electricity as that created by burning two and one-half tons of coal. Turning a small amount of material into a large amount of energy is a great accomplishment, but nuclear scientists are working on an even greater marvel, the breeder reactor.

Breeder Reactors: Scientific Miracle

If you put three pounds of nuclear fuel into a reactor and "burned it all up" to produce electricity, how much fuel would you have left?

"Why, none, of course," is the answer you would probably give to this strange question. But you would be wrong if we were talking about an amazing kind of power source called the breeder reactor. In some kinds of breeder reactors, after the three pounds of fuel was used, *there would be four pounds of fuel remaining.*

WESTINGHOUSE

A single steam generator in an atomic power plant contains 39 miles of tubing. Water heated by the nuclear reactor is pumped through the tubes, shown here being welded into place.

How Will We Meet the Energy Crisis?

How does a breeder reactor perform this feat of scientific magic? The answer is that a breeder reactor does two jobs at once. It produces power, and creates new nuclear fuel.

Is such a reactor just a drawing-board dream for the remote future? Not at all. Scientists of the AEC have already solved most of the problems of developing a reactor that creates more fuel than it uses. Their program calls for this scientific wonder to go into operation by 1984.

The theory behind the breeder reactor is quite simple. As you know, a uranium^{235} atom can fission when its nucleus absorbs a neutron. The fission reaction releases free neutrons that may, in turn, initiate other fissions. All the neutrons released, however, are not necessarily absorbed by fissionable material; some are wasted by being absorbed in the structural material of the reactor, the control elements or the coolant.

The breeder concept puts the wasted neutrons to work and exploits the characteristics of certain *fertile materials*. When the nucleus of an atom of fertile material absorbs a neutron, the fertile atom can be transformed into an atom of a *fissionable material*— a different but very desirable substance. By careful selection and arrangement of materials in the reactor —including, of course, fissionable and fertile isotopes —the neutrons not needed to sustain the fission chain reaction can convert fertile material into fissionable material.

WESTINGHOUSE

This component of a coolant pump for an atomic reactor was carved from a single piece of stainless steel. As it moves 93,000 gallons of water a minute through the reactor, it has to be strong to withstand the pressure and heat.

How Will We Meet the Energy Crisis?

If, for each atom that fissions, more than one atom of fertile material becomes fissionable material, the reactor is said to be breeding. One fertile material is uranium^{238}, which is always found naturally with fissionable uranium^{235}. When uranium^{238} nuclei absorb neutrons, they are converted to nuclei of fissionable plutonium^{239}.

Present-day reactors, which just "burn up" atomic fuel, manage to use only about 1 per cent of the energy locked within uranium ore. With a breeder reactor it may be possible to use as much as 50 per cent of the energy in uranium. In other words, it will take far less uranium to do the same amount of work.

The consequences of this enormous jump in energy utilization are outlined by Dr. Glenn Seaborg, Chairman of the U.S. Atomic Energy Commission:

Over the next 50 years, use of breeders as planned can be expected to reduce by 1,200,000 tons the amount of uranium that would be consumed without breeders. That is the energy equivalent of about 3,000,000,000 tons of coal.

The present cost of producing electricity in the U.S. ranges from five to 10 mills per kilowatt-hour delivered to the transmission system, depending on the type, age and location of the plant. This range covers most plants, although there are a few outside of either extreme. The liquid-metal fast breeder reactor is predicted to produce power at a saving of from 0.5 to one mill per kilowatt-hour. Large breeder-reactor systems that eventually bring the cost of electricity down as much as two mills per kilowatt-hour will make it possible to extract, use and reuse resources in ways that cannot be afforded today. It will be possible to tap substantial resources in the oceans and on land and to use land not now habitable or productive.

Indeed, we believe breeders will result in a transition to the massive use of nuclear energy in a new economic and technological framework. The transition may be slow, and it will require the introduction of a series of innovations in the technologies of industry, agriculture and transportation. The innovations will include large-scale, dual-purpose desalting plants, electromechanization of farms and of means of transportation, electrification of the metal and chemical industries and more effective means for utilizing wastes. The key to these possibilities is abundant low-cost electrical energy, and the route to that is by way of the breeder reactor.

Fusion—Power for Tomorrow

Even the vast potential of breeder reactors falls short of the dazzling possibilities that lie in fusion power. The energy that could be released by joining atoms together, rather than splitting them apart, staggers the imagination.

How Will We Meet the Energy Crisis?

Fusion is the process which gives our sun its energy. When hydrogen atoms join, as they do under conditions of tremendous heat, they fuse to form an atom of helium. In doing so, they give off some of their mass in the form of energy. In the sun, 40,000,000,000 tons of hydrogen atoms are thus fused every minute. There is so much hydrogen in the sun that even at that rate it will take billions of years to consume it all.

Our planet, too, has a huge supply of hydrogen—in seawater. For every 6,500 or so atoms of ordinary light hydrogen in water, there is one atom of deuterium—"heavy water." This deuterium could be used in a fusion reactor. Even by today's methods it costs only about four cents to extract the deuterium in a gallon of water, and this amount contains as much energy as 300 gallons of gasoline.

So scientists start out their thinking about putting fusion to work with the tantalizing fact that it could make available an almost inexhaustible supply of cheap atomic fuel. Some experts calculate that no matter how great power demands might be, there would be no chance that we could run out of fuel for literally millions of years.

What's the problem? It can be expressed in one word—heat. To get energy from the fusion process it is necessary to create heat of around *180,000,000° F.* Scientists have already created such heat in an uncontrolled atomic device—the hydrogen bomb. To

Weird looking devices like these are tools of scientists seeking ways to control plasmas that will make fusion power possible.

LAWRENCE RADIATION LABORATORY

trigger its fusion reaction, a smaller fission bomb must be exploded. Obviously, such a bomb could hardly be used in a power plant! However, the nuclear engineers have a number of different ways that might be used to produce the necessary heat. It might be done with electric currents that would vibrate hydrogen plasma, with microwaves, with strong magnetic fields or with rays from lasers.

The real stopper is: What kind of a container can you use to hold anything that hot? No metal or other material in existence, or any substance science can even visualize, could withstand 180,000,000° F. heat. The answer is not to use a material at all, but a force. Physicists believe that by using superconductive magnets (see Chapter 5) they can create "magnetic bottles." Researchers at many nuclear facilities have made much progress toward developing various kinds of magnetic containers.

Success in taming nuclear fusion may not be achieved before the year 2000. Some forecasters say it is more likely to be closer to the middle of the 21st century before this mighty force will be harnessed for mankind. However, though they may disagree as to timing, all scientists agree that it is a matter of *when,* not *whether*. They are sure that the power-hungry world of tomorrow will have to find a way to use the energy-creating process that powers the great stellar furnace, our sun.

4
HOW SAFE ARE ATOMIC POWER PLANTS?

"The city of tomorrow will be built around an atomic power plant."

This forecast, made by city planner Dr. Athelstan Spilhaus, seems frightening to many people. In spite of the 100 per cent safety record of atomic power plants up to now, many searching questions are still being asked.

Is it really safe to put atomic power plants in congested urban areas? Isn't there a possibility of dangerous radioactive fallout? How about contamination of water with radioactivity? And, above all, what about the danger of a nuclear accident? Couldn't an atomic power plant blow up, like an atomic bomb?

How Will We Meet the Energy Crisis?

The questions are pertinent enough to have aroused the concern of many citizens, including some eminent scientists. No matter how brilliant the future of atomic power may be, no matter how much we may need such plants, are we justified in building them close to cities? Groups interested in protecting the environment have gone beyond simply asking questions; they have blocked the construction of nuclear power facilities in many metropolitan areas. Yet, as the demand for power grows, the pressure to build many more nuclear plants increases steadily. An estimated 500 will be built in the United States by the year 2000.

Few scientists believe that atomic plants, in their normal day-to-day operation, present any fallout danger at all. The AEC has set up rigid requirements based on medical studies of the amount of radiation which might be harmful. Their standards start with a consideration of background atomic radiation, which is very much a part of your life wherever you happen to live.

Background radiation comes from two sources. One is radiation in the form of cosmic rays—high-energy particles from outer space. The other source is natural radioactivity—that is, that which comes from naturally radioactive substances present in commonplace materials. Part of the potassium and carbon in our bodies, for example, is radioactive.

Radioactivity is measured by the unit called a mil-

U.S. ATOMIC ENERGY COMMISSION

Experiments are constantly carried out to make reactors safer and more efficient. Research reactors such as this one are valuable for tests that help improve the efficiency and safety of nuclear power plants.

lirem, and you're exposed to anything from 90 to 200 millirems per year in natural radiation if you live in the United States. A generally accepted average figure is 125 millirems. In some parts of the world, at very high altitudes, the amount is much higher. The Federal Radiation Council and the National Committee on Radiation Protection and Measurements, two government organizations set up to set standards for nuclear safety, reached the conclusion that a safe figure for exposure to radiation is 500 millirems. This recommendation is embodied in the Code of Federal Regulations, which has the force of law. Atomic power plants, therefore, must be designed to release no more than this amount. How much do they actually release?

"None that we've been able to detect," say the atomic scientists charged with monitoring the air in the vicinity of power plants. Armed with incredibly sensitive instruments, on the ground and carried aloft by helicopters and airplanes, they report that they have never been able to detect any radiation above the normal background level for the particular area.

Some critics dispute the findings of the AEC, however. They maintain that emissions from some types of atomic reactors may be higher than those from other types. A fierce controversy has arisen, for example, about the Dresden reactor in Illinois, 50 miles southwest of Chicago. This reactor, which started up in 1959, was known to emit measurable amounts of radioactivity (although these amounts were safely within prescribed limits). Dr. Ernest J. Sternglass, professor of radiation physics at the University of Pittsburgh School of Medicine, maintained that this radiation had produced a definite effect on infant mortality in areas near the plant. His figures were disputed by Illinois public health authorities.

Similar controversies have arisen over the possible contamination of rivers and lakes with radioactive substances. In these cases, there seems to be little cause for alarm. Quantities of water are used for cooling in atomic plants, just as in fossil fuel plants, but the only water that goes from a waterway into the plant and then empties back directly into the waterway is that which is used to cool the turbine condensers, which are not radioactive. This water does not flow directly through the reactor.

Water used in the reactor itself is put through a long series of filtering and evaporative processes. Only after almost all radioactivity has been removed is the water returned to the stream. In a typical atomic power plant the amount of radioactive substance thus returned amounts to only a few millionths of a gram. The water is perfectly safe to drink.

Some ecologists, however, are not sure that the tiny amount of radioactivity released is as harmless as AEC scientists contend. It has been found that in

PHILADELPHIA ELECTRIC CO.

Many nuclear power plants, like this one near Philadelphia, have been built in urban areas.

Scientists use complicated monitors in an airplane to measure radioactivity in the atmosphere. There is only the remotest possibility that any would ever escape from a nuclear power plant.

E G & G, INC.

some waters certain species of fish tend to build up radioactivity in their bodies in concentrations thousands of times greater than the amount of radioactivity in the water in which they swim.

One ecological report states that "scrutiny of wildlife in a pond receiving runoff from the Savannah River Plant near Aiken, South Carolina, disclosed that while the water in that pond contained only infinitesimal traces of radioactive zinc^{65}, the algae that lived in the water had concentrated the isotope by nearly 6,000 times and the bones of bluegills, omnivorous fish that feed both on algae and on algae-eating fish, showed concentrations more than 8,200 times higher than the amount found in the water.

"Study of the Columbia River, on which the Hanford, Washington, reactor is located, showed that the radioactivity of the river plankton was 2,000 times greater than that of the water while the radioactivity of the fish and ducks feeding on the plankton was 15,000 and 40,000 times greater, respectively. Moreover, the radioactivity of the egg yolks of water birds contained more than a million times the radioactivity in the water."

The possibility that some of the organisms building up these large amounts of radioactivity might end up as human food is worrisome. To avoid this danger, nuclear plants of the future will probably be required to return no water to waterways, no matter how small the amount of radioactivity they contain.

The most frightening question about atomic power plant safety, of course, concerns the chance of an accident. In spite of the many safeguards that have been built into nuclear power plants, some experts believe there is still a chance of disaster. Dr. Ralph E. Lapp, a noted physicist who worked on the original atomic bomb, is one of the authorities who concludes that these power facilities should not be located close to cities. He states flatly: "Before the year 2000 we will probably have 500 nuclear power reactors of 1,000,000-kilowatt rating, and it would appear a certainty that we will have a serious nuclear accident."

Many scientists will agree that with that many power plants in operation an accident might indeed occur. However, they maintain that there would be so many built-in safeguards that no damage would be suffered by the environment or the people living near the plant. Even if the plant was beset by a major natural catastrophe—an earthquake, flood, tornado or hurricane—an emergency safety device would operate. This final safety measure is called the "vapor containment system." This is the way the AEC explains it:

If you were the designer, you would begin by imagining what is usually referred to as "maximum cred-

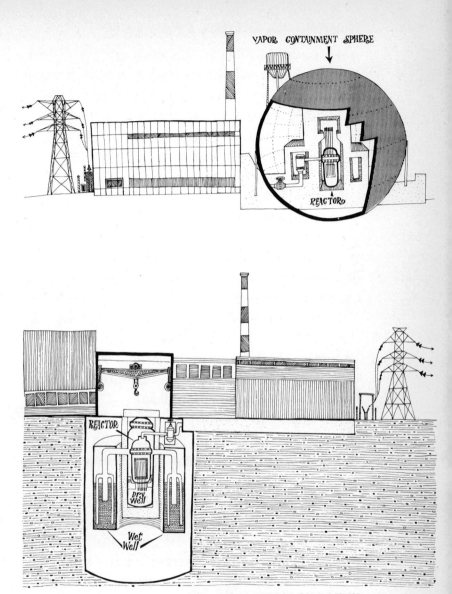

U.S. ATOMIC ENERGY COMMISSION

Two systems designed to prevent harmful radiation from spreading in the unlikely event of an atomic power plant accident.

ible accident"—that is, the most serious reactor accident that could be expected to happen if major design safeguards failed. This would involve hypothesizing not one but a combination of several highly improbable things going wrong simultaneously. Then, taking into account the size and design characteristics of the reactor, you would make some calculations.

Let us assume, by way of example, that the "maximum credible accident" involves a sudden escape of all of the coolant contained in the primary system of a water-cooled reactor—which would happen if the walls of the high-pressure system were breached. This would mean that all of the energy normally "stored" (as heat) in the coolant might be released. Picture, in short, an event similar to a boiler rupture in which a large amount of high-pressure, high-temperature water flashes to steam.

Let us also assume that the standby core-cooling system does not function properly, with the result that fuel elements overheat and cladding failures occur, releasing fission products.

In analyzing the consequences of such an accident (or, more accurately, this combination of accidents), you would first calculate the maximum pressure that could be exerted on the walls of a containment enclosure. Then, after estimating the amount of various specific radioactive substances that might be released by the overheated fuel and

How Safe Are Atomic Power Plants?

the possible rate of leakage of vapor out of the containment enclosure, you would calculate the maximum rate at which these substances could be expected to escape from the plant.

Then, taking into account the characteristics of the proposed reactor site—in particular its meteorology (prevailing winds, etc.) and its location in relation to the surrounding population—you would estimate the maximum radiation exposure that might be received by persons at the plant boundary and at outlying distances if this hypothetical series of events actually occurred.

If you found that the exposure pattern is consistent with the Atomic Energy Commission's radiation protection standards and related siting criteria, you would then be ready to proceed with the design of the containment system.

In the design and construction of the vapor containment system, the rule of conservatism applies. For example, the structural design of a containment enclosure is customarily based on a pressure higher than the calculated pressure. Wherever pipes or ventilation ducts penetrate the enclosure, precautions are taken to ensure that they do not compromise the integrity of the containment system. The same applies to access doors for personnel. When completed, and at intervals during the life of the plant, the enclosure is carefully inspected and tested to determine that it meets the degree of leak tight-

How Will We Meet the Energy Crisis?

ness specified by the design. Beyond these and other standard safeguards, special safeguards may be provided in particular circumstances.

Two principal types of vapor containment systems have been used to date in central-station plants employing water-cooled reactors.

One type makes use of a large spherical or cylindrical steel shell that encloses essentially the entire reactor installation. The shell, which in a large plant might be the height of a 20-story building, is constructed by welding together sections of steel plate. In the plants which are located at a distance from population centers, a single containment shell is used. Plants in or near population centers are built

The steel containment sphere is the dome-like structure in the center of the giant San Onofre, California, atomic power plant.

U.S. ATOMIC ENERGY COMMISSION

with a double-walled "zero leakage shell" surrounded by a massive concrete radiation shield.

A second, basically different type of vapor containment system has come into use recently. It is known as the "pressure suppression system." In one version of this system, the reactor vessel is located in a steel containment tank surrounded by a concrete radiation shield. The containment tank, termed the dry-well, is connected by pipes to a second tank, termed the wet-well, that is partially filled with water. The entire installation is housed below ground level within a building of special construction. In the event of a vapor release from the reactor, the vapor would pass into the dry-well and from there would be immediately relieved by vapor condensation; moreover, in bubbling through the water in the wet-well, the vapor would be scrubbed essentially free of solid radioactive particles.

How safe are atomic power plants? Layman and scientist alike, we'll probably continue to worry about building nuclear facilities in urban areas and about what would happen if an atomic plant located on a major fault, such as the San Andreas in California, were subjected to an earthquake.

We can only hope the scientific answers will be reassuring. It will be hard to provide enough electricity for tomorrow's world without tapping the energy locked in the atom.

5
NEW WAYS TO DELIVER POWER

In thousands of homes in New York, New Jersey and Pennsylvania, lights flickered, appliances slowed and TV pictures narrowed. Another blackout? No one even had occasion to ask that fearful question, for none of these signs that something had gone wrong lasted long enough to cause alarm. Though a whole series of giant generators in Canada and the northeastern states had broken down, electricity to replace them instantly poured in from midwestern power plants more than a thousand miles away.

By the 1980s few people will even remember the days when there were blackouts due to breakdowns of generators. If a power source anywhere in the United States should fail, the current lost will not

New Ways to Deliver Power

create an emergency. New transmission systems will make it possible to deliver power over vast distances —from the West Coast all the way to the East Coast, for example. The United States will be covered by one huge power grid, as will Europe and large parts of other continents as well.

Of course, inter-ties—systems that link one power system to another—are commonplace today. Much of the northeastern and midwestern parts of the United States are already linked. Increasingly, there is a need to transfer power from one area to another as demands change or vary from region to region. The big difference will come in the form of radically new methods of power transmission that will overcome the two big problems that confront the power producers: How can they send enough power along any one transmission line? How can they prevent power losses?

More Power on the High Towers

The transmission towers marching across the land like metal giants from another planet will probably disappear from the landscape as new systems replace them. However, these overhead high-voltage lines are what power men have to work with now, and they're making great strides toward getting them to do far more work than they have ever done. With projects like Churchill Dam and the coal-fired plants of WEST being built, engineers of the early '70s have been faced with the task of delivering a torrent of power over the kind of cables they have today.

Their key tool is a giant version of the familiar black box—the transformer you see atop ordinary residential power poles. A transformer is a device that steps up—or steps down—voltage. In the case of power to be delivered over a distance, the power engineers want to step it up—way up. The higher the voltage, the more electricity they can push along a wire. Current, as it comes from the generators, wouldn't get very far because it's just not pushed out with enough force—enough voltage, in other words. Even a powerful generator may be developing only 20,000 volts. To build up a voltage that would carry current for a hundred miles or more, it would be necessary to go up to 100,000 volts. For a long time, this seemed about the highest figure that could be reached. However, by the early 1960s the ingenious technologists had found new materials and new ways to wind the miles of wire that go into a big transformer. They were up to an incredible 345,000 volts.

Even that was not the limit. In 1968, Westinghouse engineers triumphantly unveiled a giant transformer that took over where the most powerful previous ones left off. A metal monster 50 feet high, 21 feet wide and weighing over half a million

AMERICAN ELECTRIC POWER SERVICE CORP.

Transmission towers require a wide right-of-way that uses up large areas of land.

SOUTHERN CALIFORNIA EDISON CO.

Power to feed electricity-hungry cities flows along new, high-voltage transmission systems.

How Will We Meet the Energy Crisis?

pounds, it steps up voltage to 765,000 volts. Installed with others like it in a five-state network of transmission lines of the American Electric Power Service Corporation in Ohio and neighboring states, it enables these lines to carry *five times* as much power as they did when operating at 345,000 volts. Since then engineers have designed transformers that boost currents to well past the million-volt mark.

Supercold—Electricity Carrier for Tomorrow

These advances certainly make an important contribution to power distribution, but they're only a step along the way toward transmission systems that can meet the enormous demands of the future. One thing seriously wrong with present systems is that they lose so much valuable electricity. Losses along the line, and in the process of going through the transformers, can amount to 20 per cent. Even though losses are less at high voltage than at low voltage, they put a serious dent in the amount of electricity delivered.

Scientists are sure they have found a way to deliver electricty with *no* loss. The story began in 1911, when Heike Kamerlingh Onnes, a Dutch physicist who was later awarded the Nobel prize, made an astonishing discovery. Experimenting with the use of liquid helium to push temperatures down to

AMERICAN ELECTRIC POWER SERVICE CORP.

A huge substation that controls and regulates the 765,000-volt current which moves along a network of power lines.

something close to absolute zero (minus 460° F.), he tried sending electric current through supercooled copper wires. To his amazement, meters showed something that science had always considered impossible. All materials—even copper, which is a good conductor—have a certain amount of resistance to the passage of electricity through them. Yet Kamerlingh's instruments showed no resistance whatsoever.

Repeated experiments produced the same results. The physicist drew the conclusion, accepted by the scientific world, that when certain metals are cooled to temperatures near absolute zero, they become perfect conductors. They carry current without heat or energy loss of any kind. Engineers realized at once the enormous possibilities opened up by this discovery of superconductivity. Find a way to cool electrical wiring in motors, generators, electromagnets, transformers and transmission lines, and a whole new world of electricity would open up, because of the efficiency possible in devices that lost no energy.

The scientific dream remained just that for more than 40 years, however. As researchers plunged into investigations of the phenomenon that looked so promising, they made a dismaying discovery. They found plenty of metals—mercury, lead and tin, as well as copper—that were superconductive when

New Ways to Deliver Power

cooled. The trouble was that when any large amount of current was introduced into frigid metals, a magnetic field came into being around them; the more current, the stronger the magnetic field. When the field got strong enough, it canceled out the superconductive effect. The metals reverted to their natural state of limited conductivity.

The unhappy researchers, thus cheated of any practical use for superconductivity, did not give up their dream. In Europe and the United States, many scientists continued to investigate this strange phenomenon, driven by the hope that someday they might find some material that would not be affected by those destructive magnetic fields. Decades went by, with little progress. Then, in 1953, physicist Bernd Teo Mathias and a team of researchers at Bell Laboratories made a breakthrough. They found that an obscure gray metal, niobium, and a related compound, niobium-tin, kept their conductivity even when powerful currents were shot through them.

Many problems lay ahead. For one thing, niobium is brittle and hard to work with. But, one by one, difficulties were overcome as ways were found to combine niobium with other metals. By the mid-1960s the superconductor was actually put to work in a new breed of electromagnets. Once current is introduced into these magnets, it keeps moving around and around the coils. Engineers calculate that in one magnet the original current would not run down for 20,000 years!

When scientists at Argonne National Laboratory near Chicago needed a giant electromagnet, they found that operating an ordinary one would require 10,000,000 watts—enough to power a medium-size city. Instead they built a superconducting magnet. Cooled by liquid helium, it requires only 500,000 watts—one twentieth of the amount needed by an ordinary magnet. It saves this government-operated research laboratory nearly $400,000 a year in electricity bills.

While such magnets demonstrate the possibilities of deep cold, they are only a step along the way toward putting supercooling to work in transmitting electricity. The systems that engineers have worked out would probably run cables through underground pipes filled with liquid helium. A supercooled system could not economically be used above ground. This fact is not a disadvantage, however, because of the way it ties in with another pressing problem that confronts power engineers.

"We'll have to go underground," says the head of one of America's biggest utilities. "If, for no other reason, because the environmentalists aren't going to let us stay above ground."

The marching transmission towers that once seemed symbols of progress are now the targets of many individuals and organizations who maintain that these steel monstrosities scar the landscape. Even if they were esthetic, they use up a staggering amount of land. For example, a proposed 120-mile-long, 345,000-volt transmission line to run between Ramapo and Binghamton, New York, would absorb 3,000 acres of land—five square miles.

The New York Times was prompted to ask in an editorial, "Should the countryside be butchered in order to give city dwellers air conditioning?"

Utility company officials agree that the power lines are unsightly and that they are land grabbers, but they point out that to get all transmission lines out of sight, underground, would cost as least $350,-000,000,000! Many experts set the figure much higher.

That's where superconductivity enters the picture. It could cut the cost of burying power lines to a fraction of what it would otherwise be. The capacity of underground lines would be tremendously increased—so much so that one authority, Dr. Bruce C. Netschert, of National Economic Research Associates, Inc., confidently estimates that "one-third of the total power requirements of New York City could be supplied through a single cryogenic cable inside a pipe 18 inches in diameter."

The obstacles in the path of underground superconducting are numerous, but researchers are hopeful that they will find a way to use inexpensive liquid hydrogen instead of liquid helium. This would greatly reduce costs—to perhaps as little as a hundredth of those of using liquid helium. No one can make exact predictions as to exactly when supercold can take powerlines underground, but some forecast that by the mid-1980s unsightly, power-losing overhead transmission systems could be a thing of the past.

Power Through the Air

Looking still farther into the future, scientists foresee an even more revolutionary way of transmitting electricity. They propose to send power through the air! This seemingly wild idea, first proposed by the electrical genius Nikola Tesla, is no longer branded as a visionary scheme. On a small scale, its workability has already been proved. In a demonstration at the laboratories of the Raytheon Company in Burlington, Mass., visiting scientists, reporters and Air Force officials actually saw the beginnings of a new age of power transmission. As they watched, a model helicopter with no engine or batteries rose from the ground. The electric motor that turned its rotors was receiving its power in the form of microwaves hurled through the air.

To be sure, the helicopter was just a model, and the amount of power delivered was not great, but the three basic components of a microwave power

Engineers preparing for a demonstration of what may become the transmission system of the future—power sent through the air! Receivers in the miniature helicopter model pick up microwaves, turning them into electricity which powers a motor which turns the rotors.

RAYTHEON CO.

Up it goes! Invisible energy sent through the air takes the helicopter aloft. The cables are used as guidewires.

RAYTHEON CO.

RAYTHEON CO.

Behind the scenes is the apparatus that turns electricity into microwaves that can be beamed through the air. Such experimental equipment is not very big or efficient, but it shows the possibilities of microwave transmission of power.

transmission system were all there. There was a device that turned electric power into microwaves, another device that beamed them at the helicopter and a receiver on the helicopter to pick them up and convert them back to the electricity that ran the motor.

The Raytheon engineers credited their breakthrough to advances made in other technologies. The powerful tube that produced microwaves was an adaptation of tubes developed for use in communications, such as the microwave relay systems that carry television programs. The device that hurled them toward the target was an outgrowth of the giant antennae used by radio astronomers to send questing radio beams into the depths of space. The key device aboard the helicopter, called a "rectenna," was a combination of radar-type antennae and rectifiers.

It may be a long time before full-scale transmission of power without wires becomes a practical reality. Yet, with power needs going steadily up as the world nears the 21st century, it may prove to be the best way to deliver the electricity simply and inexpensively without damage to the environment. And it may turn out to be the tool that enables us to capture the ultimate source of power abundance —the energy of the sun.

6
MINIATURE POWER PLANTS

The black funnel of a tornado roars across the midwestern countryside, cuts a swath of destruction through a city, then thunders on beyond it to commit a final act of violence. As if they were toys, it topples a row of transmission towers.

With night coming on and a city in ruins facing a blackout, harried officials send out a call for help.

"We need a power station in a hurry!"

Within hours, a fleet of huge trucks has rolled into the stricken community to cluster around a local power substation. Soon electricity is flowing again. A miniature portable power plant has come to the rescue.

Earthquakes, floods, hurricanes and tornadoes. On

Miniature Power Plants

this uncertain planet of ours there is often need for electricity that cannot be delivered by the usual transmission lines. A weather station in Greenland, a space station orbiting the earth, a base on the moon—these, too, are situations where it just isn't possible simply to plug in to get power.

At times and places like these, the power plant itself must be taken to where the electricity is needed. Engineers have come up with many different approaches to portable power, and are even looking into ways of putting a "power plant in every building." New developments point to many possibilities for achieving truly "portable electricity" in tomorrow's world.

Portable atomic power plants can provide electricity in emergencies—earthquakes, tornadoes, floods, hurricanes.

U.S. ATOMIC ENERGY COMMISSION

How Will We Meet the Energy Crisis?

Portable Atomic Reactors

A city under the ice!

That was the astonishing proposal made by scientists who wanted to carry out a variety of research programs in the Arctic.

When scientists of the U.S. Navy decided to build a research city deep under the ice cap in Greenland, they issued a challenge to nuclear engineers. For Camp Century, as the experimental base was to be called, they would need a source of power to light the thousands of feet of tunnels and the many rooms they proposed to cut out of the ice. Was it possible to build an atomic reactor small enough to be transported to this remote spot and safe enough to operate close to where the scientists and other personnel would be living and working? To be sure, they could use diesel engines to operate generators. However, if they put such engines under the ice, fumes would be a problem, and if they put them above the ice, they would be subjected to all the fury of the Arctic weather. Moreover, the cost of oil delivered to this ice-bound outpost ran as high as $6 a gallon.

The atomic scientists of the AEC and its subcontractors were ready to meet the challenge. They devised a reactor that could be dismantled. Its parts were hauled by ship, plane and snowmobile and put back together again in a room hollowed out of the ice. The transported nuclear reactor continued to supply all the electricity needed for Camp Century —as much as that needed for a city of 20,000 people.

This experimental unit of 1960 paved the way for many similar reactors designed for use in special situations. For example, there was the problem that confronted engineers charged with the task of building a radar station far from any power line, high atop a mountain near Sundance, Wyoming. How would they provide the large amount of needed electricity? A nuclear reactor was built in 27 different parts, none of which weighed over the 30,000 pounds that could be handled by plane and helicopter. Not one of the parts was more than 8 feet 8 inches square or more than 30 feet long.

In the future, power experts visualize moving such plants into areas in which power plants have been seriously damaged or where power lines from outside may be down for a long time. The reactors could be quickly activated and would stay in operation for as long a time as needed.

In such stricken areas, cut off from the outside world, these reactors might even provide fuel to operate motor vehicles. They could become an "energy depot," literally making a fuel that would operate internal combustion engines.

Scientists have long known that, with enough electricity, it is possible to make combustion fuels from such unlikely sources as air and water. With

U.S. ATOMIC ENERGY COMMISSION

A component of a miniature atomic reactor, bound for a destination in the Arctic, is loaded aboard a plane.

electricity, hydrogen can be taken from ordinary water and combined with nitrogen in the air to produce ammonia—which can be burned in specially designed engines. Brought into a fuel-short area, a mobile nuclear power plant could produce ammonia or some other gasoline substitute.

Atomic Energy in Small Packages

When the Apollo 12 astronauts ended their stay on the moon in November, 1969, they left behind them an array of electrically operated instruments which were intended to send back information long after the departure of the men who had left them there. Where did the electricity come from? Its source was a tiny atomic power plant using radioisotopes.

This generator was still faithfully powering the instruments when the Apollo 14 astronauts made their moon landing, at a different location, in February, 1971. They too left behind them a miniature atomic power plant which, scientists estimate, will keep operating for years.

The tiny radioisotopic generators bear the name SNAP—which stands for Systems for Nuclear Auxiliary Power. The first, known as SNAP-3, was first activated experimentally in 1959 when President Eisenhower pushed a button on his desk at the White House. It actuated a grapefruit-size generator which, small though it was, could deliver 11,600 watt hours of electrical power.

SNAP-3 operated on the same principles as all its successors. Inside this pioneering generator, a fuel supply of unstable polonium^{210} decayed, emitting high-speed particles. When these particles collided with atoms in a shell around the radioactive material, they were absorbed by the material in the shell. The collisions caused the atoms of the shell material to move more rapidly, raising the temperature inside the generator. In SNAP-3, this heat was conducted to thermocouples, and electricity was produced—a steady 2.5-watt power supply.

The nuclear engineers can make a SNAP generator to order by using a different fuel. They have their choice of eight isotopes, each one useful for particular jobs. Cerium^{144}, for instance, has a "short" half life of less than a year. It also has a high "power density" (which means simply that it can produce a lot of heat energy per pound of weight). Thus cerium^{144} radioisotopic generators would work well when you wanted a generator that put out a lot of electricity per pound but did not need a generator that would run for a long period of time.

Plutonium^{238}, on the other hand, has a half life of 89 years. Generators built with it would be better if you wanted a longer-lasting source of electricity.

The other fuels on the list of eight have other advantages. Promethium^{147}, for instance, has an ex-

Miniature Power Plants

tremely high melting point—1,300° C.—making it valuable in applications where a generator might have to withstand high temperatures.

Strontium^{90} as a fuel gave the scientists a real headache, for it is a threat to living creatures. It is a "bone-seeking" substance; that is, if taken into the body of a living creature it tends to accumulate in bones, causing damage to the organism.

When the problem of finding a way to use strontium^{90} without danger was turned over to the chemists at the Martin Company laboratories, they met, at first, with discouraging results. But finally they found that a compound chemical, strontium titanate, could be manufactured. Stronium titanate melts only at extremely high temperatures. Generators using this fuel form would be safe in fires. Further, strontium titanate does not dissolve easily in either fresh or salt water. Thus it would not be released in dangerous quantities into an ocean or other body of water. Today, strontium^{90} generators are one of the most useful kinds of radioisotopic power sources.

The SNAP generators have a long history of service in space. Soon after SNAP-3 proved that radioisotopic generators could work, the SNAP program was in high gear. A few months later, SNAP-3A roared into orbit aboard a rocket from Cape Kennedy. This first space-going generator was roughly spherical in shape and approximately 5 inches in diameter. The fuel, plutonium^{238}, was contained in a

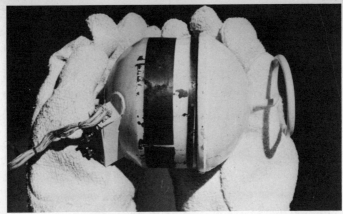

MOUND LABORATORY, AEC

One of the earliest applications of miniature atomic power plants was this SNAP unit, used in a navigational satellite sent aloft in 1961. Isotopic-decay heat makes it too hot to handle without asbestos gloves.

A SNAP-27 atomic power plant is carried by craft landing on the moon. The power unit rides in the capsule outside the landing craft; other elements of the reactor are in the compartment at the front.

NASA

Miniature Power Plants

rugged capsule inside the sphere. Mounted on the outside of the rocket, this space generator put out 2.7 watts of electricity, which was used to power the satellite's radio transmitter. Years after launch, the generator was still working, demonstrating the reliability of this new kind of power source in space missions.

SNAP-9A, an improved version of SNAP-3A, was ready for space a short two years later. These generators were nearly ten times as powerful as their predecessors. In 1963, two satellites were orbited that drew their entire electrical supply from the SNAP-9As. The less powerful SNAP-3A had been used only to supplement power delivered by solar cells.

SNAP radioisotopic generators in space are helping increase our knowledge of world weather. SNAP-19 generators, fueled with plutonium^{268}, ride in the *Nimbus* weather satellites. These orbiting weather watchers circle the earth at altitudes of 500-600 nautical miles. They constantly photograph world weather patterns and relay pictures to stations on earth. In the *Nimbus-B* satellite, two of these 30-watt generators are hooked together to supply a total of 60 watts of power.

While the remarkable series of SNAP generators is making history in space, we are also finding many down-to-earth applications for these versatile packaged power sources. At lonely, uninhabited Axel

How Will We Meet the Energy Crisis?

Heiberg Island, 700 miles from the North Pole, scientists and technicians uncrated and assembled the world's first atomic-powered weather station. It was designed to collect data on temperature, wind and barometric pressure in the Arctic, where much of the Northern Hemisphere's weather is made. A radio transmitter sent out this information every three hours.

Powering the transmitter was a SNAP generator using strontium^{90} as a fuel. Even under bleak Arctic conditions, this generator produced a constant five watts of power. This electricity was stored in rechargeable chemical batteries between transmission times. Heat from the generator was also used to keep sensitive components in the data-gathering system from getting too cold. The Axel Heiberg weather station continued to relay its important weather information for two years. Another SNAP-powered weather station has operated at Minna Bluff, 700 miles from the South Pole.

Other untended weather stations have been built to send data from far out at sea. The U.S. Navy has long maintained a number of floating barges, anchored at sea with weather-watching instruments aboard. Once these barges were powered with ordinary batteries, but recently the Navy has begun using SNAP-7D generators. These 60-watt generators use strontium^{90} fuel and will continue to supply electricity for years without refueling.

Radioisotopic generators can power navigational aids, such as buoys, beacons and lighthouses. The SNAP generators used in navigational buoys weigh nearly 700 pounds less than the chemical batteries that once powered these lonely ocean outposts.

Thermoelectricity—New-Old Path to Compact Power

Put two pieces of metal together, and heat one of them. What happens? An electric current begins to flow and keeps flowing as long as one metal is hotter than the other.

This was the discovery that astonished an early German scientist, Thomas Johann Seebeck, tinkering in his laboratory in 1821. The trouble was that Seebeck didn't know what he had discovered. If he had, the whole history of applied electricity might have changed. The world might have had a dependable source of electricity a good 50 years before generators came into use.

The experiment that Seebeck carried out was a simple one. When he applied heat to a circuit constructed of two different conducting materials, a magnetic needle held near these pieces of metal deflected from true north. The puzzled physicist didn't stay puzzled long enough. He jumped to the conclusion that differences in temperature somehow created magnetism in the metals. This was the theory

PHILLIPS PETROLEUM CO.

A technician installs a miniature atomic power plant on an offshore oil and gas platform.

Working with manipulators behind three feet of leaded glass, technicians load a fuel capsule into a miniature electric power plant—a SNAP 7-B generator.

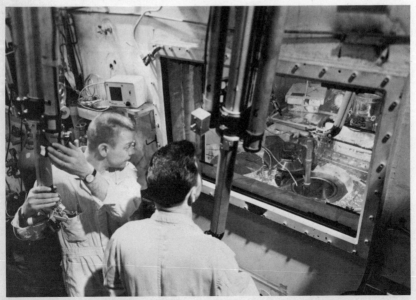

U.S. ATOMIC ENERGY COMMISSION

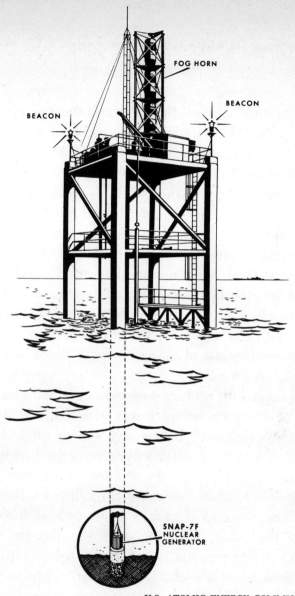

Ships won't collide with this offshore oil platform. They are warned away by the nuclear operated beacons and fog horn.

he announced to the world, pointing out that the earth itself was a good example. Differences in temperature between the poles and the equator, he said, explained the earth's magnetic field.

A few other scientists expressed the idea that just possibly an electric current, rather than magnetism, had been induced in the metals. Seebeck scoffed at such a notion. Experiments carried out with conducting metals, such as copper, seemed to prove him right, because researchers could detect no measurable current. What neither they nor Seebeck realized was the fact that what we now know as the thermoelectric, or Seebeck, effect doesn't work very well with conductors. It works best with semiconductors—the class of materials from which today's transistors are made.

Though 19th-century scientists eventually found that Seebeck was wrong, it was more than a century after his discovery before any serious effort was made to put thermoelectricity to work. By that time technologists knew more about semiconductors and were able to make thermoelectric cells with them.

When one end of such a device is heated more than the other, electrons tend to move to the cooler end. In time, the hot end of the semiconductor becomes positively charged in relationship to the cooler end. In other semiconducting materials, the moving charges are positively charged "holes" left by electrons. These migrate to one end of the semi-

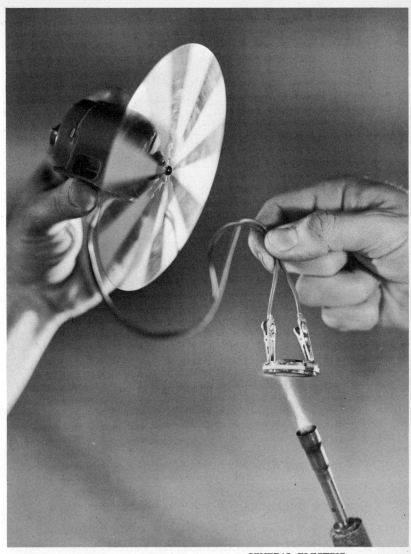

GENERAL ELECTRIC

A tiny experimental thermionic converter is turning heat into electricity which powers the fan.

conductor, producing the same kind of difference in charge between ends.

Electricity is generated with semiconductors and heat by the simple process of joining an n-type material and a p-type material together at their heated ends ("n" means that the material has an excess of negative charges; "p" that it has an excess of positive charges). If a conductor is placed between the cold ends and an electricity-using device hooked into the circuit, electricity will flow. It won't be much current if only one such circuit is being used to produce it, but when hundreds or thousands of thermoelectric cells are joined together, the strength of the current is greatly increased.

Researchers in various countries are working to improve the efficiency of thermoelectric cells. In the United States they have been put to practical use to operate furnace fans with electricity produced by gas heat. They have been installed in marine navigational markers to provide the electricity for flashing beacons. Refrigerators using kerosene burners to provide the heat are available for use in remote spots beyond the power lines. In the U.S.S.R., where thermoelectricity is more widely used than anywhere else in the world, many appliances get their electricity from thermoelectric cells. One of the most common devices is a radio receiver operated by a kerosene burner.

"Imagine," wrote Abram Joffe, a Soviet scientist

who helped develop it, "someone living in the far north of our country to whose dwelling such a device is brought. Although the snows and the tundra separate him from the rest of humanity, he is suddenly able to use a radio to hear the news of the day and music and to learn about the life of his country."

Another kind of miniature power plant that can turn heat into electricity is the thermionic converter. It looks somewhat like the vacuum tube in a radio or TV set, but the circuitry inside the thermionic tube produces electricity rather than using it. This is accomplished by "the Edison effect": When a metal is heated in a vacuum or near vacuum, it

A researcher in thermionics studies an experimental system that utilizes temperature differences to produce electricity.

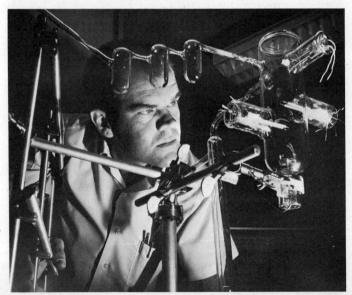

SANDIA LABORATORIES

generates electrons which will leap to a nearby colder surface.

Early thermionic converters had flat-plate positive and negative electrodes facing each other across a very short space. By heating one electrode to about 800° F. and the other to 300° F., researchers found that they could convert heat into electricity with an efficiency of less than 6 per cent.

Today's thermionic converters have their electrodes surrounded by cesium vapor. At operating temperatures, this vapor ionizes. Cesium vapor converters turn heat into electricity with as much as 17 per cent efficiency.

A Power Plant in Every Home?

The possibility of putting a power plant in every home is one seriously discussed by power technologists. It could be done, they believe, with a chemical fuel cell generator, no bigger than a suitcase, which would provide all the electricity needed by a household at less than the cost of producing it at a distant power plant and transporting it by wire.

Fuel cells have been widely discussed as a possible means of providing electricity for pollution-free automobiles of tomorrow, and that is most likely to be their first wide use. However, many experts believe they hold promise as home power producers.

A fuel cell power plant. This device generates 1,000 watts.

How Will We Meet the Energy Crisis?

Like many other power developments that may be used in the future, the fuel cell has a long past. As early as 1802, the great English chemist Sir Humphrey Davy demonstrated this way of using chemical reactions to produce electricity. A fuel cell may be described as a kind of storage battery which never needs recharging. It consists essentially of two plates, or electrodes, immersed in a chemical solution. In one typical cell, for example, this solution consists of potassium hydroxide. The electrodes are coated with another substance which acts as a catalyst. Air flows into one electrode, and fuel gas flows into the other. The energy released by the reaction between oxygen and fuel gas is converted directly into electricity. The reaction also creates water, which is drawn off as a waste product.

Fuel cells can use a variety of fuels, ranging from hydrogen to bacteria. The most likely fuel for packaged power plant use would be methane. The fuel cell would be quite small, but a number of them could be linked together to create a power plant of large capacity. This does not seem to be a likely use for cells, however. Researchers of the Institute of Gas Technology are at work on the much more fruitful idea of using them to provide home power plants.

The power plant using small fuel cells may have an economic advantage over the transmission of electricity through cables. Francis Bacon, an Eng-

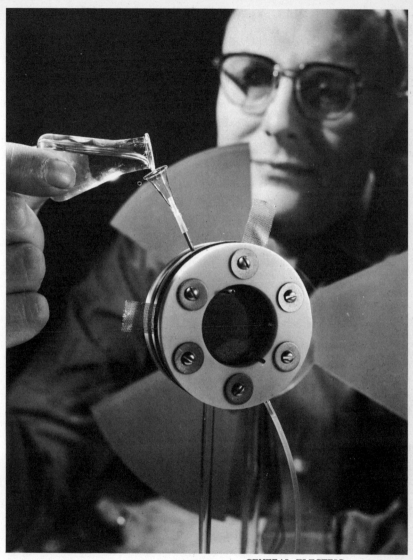

GENERAL ELECTRIC

Oil poured into this fuel cell produces electricity by chemical means, not by burning.

lish scientist who has long worked with fuel cells, points out that distributing gas by pipeline costs only one-tenth to one-fifth what it costs to distribute electricity by present-day techniques. Bacon has proposed an imaginative scheme for London and other big cities of the world. He envisions a gas pipeline network carrying liquid fuel to fuel cell units located in individual homes or small neighborhood substations. Perhaps the low cost of distributing the gas could offset the higher cost of fuel cells.

Most authorities believe that packaged power units which can be taken to the place where they're needed will never provide a large percentage of electricity. But many other experts are not so sure. They think these small-scale power plants will turn out to be increasingly important in tomorrow's world as we move into remote lands, under the sea and out into space.

7
NEW POWER FROM RIVERS

The year is 1990. The maps of North America coming off the presses look like none of the previous maps of this part of the world. A great new river system is shown, branching in and running down through Alaska, Canada and the western United States and into Mexico.

No great natural cataclysmic event has changed the face of the continent. The people of Canada, the United States and Mexico have joined together to do that. In their search for new sources of water and power, they have carried out the greatest engineering project in history.

You may find it surprising that, up to the present, hydroelectric power—energy derived from falling water—has provided only a small part of our electricity. To most of us, a mighty dam is the very symbol of power production. Who hasn't heard of Hoover Dam, Grand Coulee Dam and the dams of the Tennessee Valley Authority? Yet, for all the attention they have received, these huge dams, and others like them, produced only about 15 per cent of the power used in the United States in 1971. Perhaps this percentage won't increase much in the future, but engineers are hopeful that they can capture far greater quantities of power from the latent energy of falling water.

Distant Rivers for Power

One new solution to getting more power from rivers is to go to where they are and carry the electricity to where the people need it. Even present methods of transmission make it possible to carry power long distances, and those which will be developed in the future will make long distance power grids even more feasible. This opens up the possibility of a new type of dam. In the past, almost all great dams have been multipurpose, serving flood control and irrigation needs as well as those of electricity production. In the future this will change. Dams will be built solely to provide power, without regard for the extra benefits that have always been considered necessary. They will be constructed in wilderness

CHURCHILL FALLS CORP.

A helicopter lands equipment in the remote reaches of Labrador, where a giant power project is under construction.

areas where there is no demand for either irrigation or flood control.

A pioneering venture, the Churchill River Project, being built in the early 1970s, is a striking example of the hydroelectric power projects which may add much to world energy supplies in the future. More than 150 years ago, a party of Hudson's Bay Company traders, the first Europeans to venture onto the mighty Churchill River in Labrador, found themselves floating down this broad stream with no idea of what lay ahead. Suddenly their leader, John McLean, sitting in the bow of the boat, held up his hand in warning of danger ahead.

The men began to row frantically, driving their boat toward shore. They too could hear the thunder of falling water that had alarmed McLean. As they rowed, they could feel the tug of a powerful current drawing the boat toward—they knew not what.

Exhausted from their struggles, they finally managed to get across the increasingly strong current and draw the boat up on the rocky bank. A party of them went afoot downstream to try to find the source of the sound. As they advanced, the thunder grew steadily louder. Then there it was, the awesome sight that they were the first white men to see—the mighty falls of the Churchill River.

In his journal McLean wrote a description of what happened on that long-ago day. He described the sudden narrowing of the river, from a width of several hundred yards to "a breadth of about fifty yards as it precipitates itself over the rock which forms the falls; when, still roaring and foaming, it continues its maddened course for a distance of thirty miles, pent up between walls of rock that rise sometimes to a height of three hundred feet on either side. . . . Such is the extraordinary force with which it tumbles into the abyss underneath that we felt the solid rock shake under our feet."

To the Hudson's Bay men the mighty cataract of the Churchill River in Labrador was only a hindrance to navigation. To the geologists who came to view it after its discovery in 1839, it was a natural wonder. Engineers who studied it after the age of electricity dawned could only shake their heads sadly. All that power going to waste there in the remote wilderness where nobody could possibly use it!

Today Churchill Falls is no longer an object of frustration. Taking shape is a gigantic power project which, when completed in 1976, will be the largest hydroelectric plant in North America. Its eleven mighty turbines, driven by water dropping more than 1,000 feet, will feed over 7,000,000 kilowatts of electricity into the power grids serving Canada and the northeastern United States. Compare this to the 1,250,000 kilowatts turned out by giant Hoover Dam in Arizona-Nevada.

Taking shape alongside the huge project is a re-

CHURCHILL FALLS CORP.

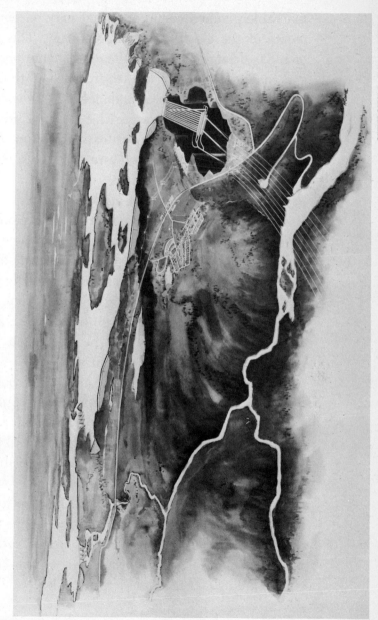

The falls of the Churchill, which once startled Hudson's Bay Company traders, are being tamed to produce electricity for northeastern Canada and the United States.

CHURCHILL FALLS CORP.

markable planned community that may become the model of the electric city of tomorrow. This town of 1,000, designed to house the permanent operating and maintenance staff, makes free use of the electricity created in such abundance by the water of the Churchill, a mile away. Electricity is used for everything—from keeping water mains from freezing in the 40-below temperature to climate control of the town center.

The center is the most remarkable feature of the community. Under one gigantic roof, it will provide a preview of the domed cities designers have long visualized. It will contain stores, offices, a hotel, an auditorium and theater, a gymnasium, bowling alley, curling rink, indoor swimming pool, a school, with grades from elementary through high school, and a public library. Buildings will be connected by a pedestrian concourse which will widen into a spacious town square. Large light traps in the roof will catch and direct light into the concourse to produce various pleasing effects of light and shadow at different times of the day. Shrubs and plants will flourish in the artificial sunlight provided by an elaborate lighting system.

Farther in the future is the grand scheme to harness rivers that flow into the Arctic Ocean, using them and various Canadian rivers farther south to provide water and electricity for large parts of the United States, Canada and Mexico. Called NAWAPA (the North American Water and Power Alliance), it is breathtaking in its scope. It would turn 40,000,000 dry acres in the United States into fertile, food-raising land and do the same for 10,000,000 more acres in Canada. It would bring water to arid Mexico, permitting irrigation there of eight times as much land as that watered by the mighty Aswan Dam project in Egypt. In addition, it would provide new supplies of water for many great North American cities, including Toronto, Los Angeles and Chicago. And, of course, it would create power—immense amounts of it.

The dream of remaking the water system of an entire continent first took shape in the early '50s in the mind of Donald McCord Baker, a water planning engineer for the city of Los Angeles. At the time, Baker had been working on plans to bring water from northern California to the dry southern part of the state. If the principle of diverting a river to supply another part of a state was workable, why couldn't the same idea be applied on a larger scale—like bringing water from rivers in Canada and Alaska south to the continental United States, and even to Mexico?

When Baker studied the map, his dream soared. What caught his attention was the Rocky Mountain Trench, a tremendous gorge that runs through the Canadian Rockies and extends into Montana. He suddenly saw this 500-mile-long trough as a storage

U.S. BUREAU OF RECLAMATION

Mighty Hoover Dam, which supplies electricity for southern California, has long been a symbol of the power that can be produced by the energy of falling water.

New Power From Rivers

reservoir. From its altitude of 3,000 feet, water could come tumbling out of the giant reservoir to drain southward.

Where would water to fill the gorge come from? There are many rivers in British Columbia and to the north, in Alaska—the Peace, the Yukon the Fraser and the Columbia, among others. Some of them drain into the Pacific Ocean, others into the Arctic Ocean, where their water is lost for man's use. What if these rivers were turned around and made to flow into the Trench?

The idea was awesome, certainly too big for one man. Its development would require the work of many engineers. Baker turned to the Ralph M. Parsons Company, an engineering firm which had worked on a variety of major projects in many parts of the world. The company engineers admitted that they were staggered by the scope of the project, but, as they explored it further, they concluded that it was a sound, practical possibility and proceeded to draw up some detailed plans for what they decided to call the North American Water and Power Alliance.

As the scheme developed over the years, the NAWAPA proposal lost none of its grandeur. As currently outlined, it calls for building a series of dams near the headwaters of the three Alaskan rivers and the Peace River in Canada. From a system of reservoirs back of these dams, water would

115

How Will We Meet the Energy Crisis?

flow into the giant artificial lake in the Rocky Mountain Trench. Some of this water would be channeled by canal through the western United States, following the mountain contours clear to Mexico, to flow into the Rio Grande and the Yaquí River. Aqueducts would branch off this main canal to serve California, Colorado, Arizona and New Mexico. An eastern branch, the Canadian–Great Lakes Seaway Canal, would flow toward the Great Lakes, reaching Lake Superior through a linking of natural lakes by canals.

The engineering problems are enormous. For instance, one of the dams will have to be 1,700 feet high. A single 80-foot-in-diameter tunnel would be 50 miles long. Building the canals calls for moving 45,000,000,000 cubic yards of earth. Structures in the project would require 70,000,000 tons of steel. A work force of several thousand men would have to toil for 10 to 20 years and the total cost might run as high as $100,000,000,000.

However, much of the cost of both construction and operation would be paid for by the enormous amounts of power produced by hydroelectric plants in the various dams. As the waters descend to sea level, they would be capable of producing enough energy to provide 30,000,000 kilowatts for Canadian use, and 38,000,000 kilowatts for the United States. Installation of additional power plants could almost triple this vast outpouring of energy, increasing it to around 180,000,000 kilowatts—the equivalent of the electricity produced by 15 Churchill River projects.

Will this colossus of engineering projects ever be built? Various governmental agencies in the United States and Canada have investigated it, and opinion is divided. One objection is that it is too expensive. Another one, raised by many Canadians, questions the desirability of flooding the immense Rocky Mountain Trench.

However, many think these objections are not serious in view of the desperate need for more water that will confront us in tomorrow's world. They point out that the $100,000,000,000 is not much more than the United States spent in a single year on its military budget during the Vietnam war. The flooding of a large area in Canada would, these supporters maintain, be offset by the recreational possibilities of such a huge body of water.

On other continents, too, engineers are at work on hydroelectric projects to capture the power of rivers. In Australia the Snowy Mountains Scheme will produce 3,740,000 kilowatts of power. The U.S.S.R. is developing many large hydroelectric projects on the rushing rivers of Siberia.

While the energy of moving water, turned to electricity, can make an important contribution to the power supply of the advanced countries, it can play a particularly significant role in the progress of

SOUTHERN CALIFORNIA EDISON

A bonus from hydroelectric power plants is the kind of recreation facilities they provide in the form of man-made lakes like this one in the high Sierra Mountains near Fresno, California.

underdeveloped areas of the world. Most of the poorer nations lie in the tropics and subtropics, which have a lot going for them in the way of water power potential. In the belt around the Equator, the hot sun evaporates huge amounts of water from the sea. Carried inland, this moisture drops as rain, giving some of the world's mightiest rivers an enormous flow of water which could be put to work producing hydroelectric power.

In Africa, especially, electricity from running water offers great opportunities for the developing nations of that continent to meet their power needs. The Congo River basin is estimated to have one-fourth of all the water power potential in the world.

New Power From Nearby Rivers

Tapping distant rivers provides a spectacular way of increasing hydroelectric power, but there is another approach that holds equal promise, particularly in the United States—increasing the power-producing capacities of nearby rivers which have already been harnessed. One good way to do that is to increase the number and size of turbines that derive their power from the water racing through the dams' penstocks.

Look what is being accomplished at the "new" Grand Coulee. This huge dam, one of the seven wonders of the power world, is famed for its mighty production of 2,200,000 kilowatts. It is about to become a far greater wonder. With the installation of a series of giant new generators, its power production is being increased by more than 7,000,000 kilowatts. The capacity of the dam when the last new generator is installed in 1974 will thus be increased to 9,200,000,000 kilowatts—making it the greatest power plant of any kind anywhere in the world.

A major engineering breakthrough will make rivers yield far more usable power than anybody thought possible just a few years ago. The trouble with falling water as a source of electricity has always been that there's no way to store alternating-current electricity. In any power system, during any 24-hour period, there are wild variations in the de-

U.S. BUREAU OF RECLAMATION

New additions to Grand Coulee Dam will make it the largest hydroelectric power plant in the world.

New Power From Rivers

mand for electricity. It is low in the early morning hours and climbs steadily during the day, reaching a peak by evening. On weekends it is low. On sweltering summer days, when air conditioners are running full speed, demands shoot up, as the citizens of many cities which have experienced "brownouts" can testify.

In steam plants, daily, seasonal and occasional fluctuations are taken care of easily. The plants can be slowed down or "fired up" to meet current needs. But with water power that can't be done. Either the dynamos are running or they aren't.

"If only we could store electricity!" has been a thought often expressed wistfully by engineers. The fact that they asked another question has led to a development that will make nearby water power sources yield far more electricity when it's needed.

The question the engineers went on to ask was: "Why not store the *water?*" The answer to that question is: Why not indeed? So they evolved the system called "pumped storage," a procedure which sounds ridiculously simple. They pump water from a low reservoir to a high one. At times when the demand for electricity is low, the generators are kept running, with the electricity they produce being used to pump water to the high reservoir. When the demand is high, the water is released from both reservoirs, powering extra generators.

With a looming electricity shortage, many power

121

WESTINGHOUSE

This diagram of the Seneca Power plant, in Pennsylvania, shows the features of a pumped storage facility: 1. Power plant 2. Main reservoir and dam 3. Storage reservoir 4. Tunnel from storage reserve to power plant. 5. Allegheny River.

New Power From Rivers

companies are rushing to complete pumped storage facilities. A large one is being built on the Grand River near Tulsa, Oklahoma; another on the Allegheny River in Pennsylvania; still another in North Carolina, near Charlotte. In countries all over the world, pumped storage systems are taking shape on the drawing boards or are actually being built.

The greatest pumped storage operation of all will begin operation in the early '70s in Massachusetts. The Northfield Mountain Hydroelectric Pumped Storage Facility is a giant installation that will provide a million kilowatts of power to electricity-short New England. And it will do it by simply making a river work harder.

The way it does it is this: It pumps water from the Connecticut River up 100 feet to a huge 300-acre reservoir excavated in Northfield Mountain. This reservoir is capable of storing 5,800,000,000 gallons of water. To construct it, it was first necessary to blast a large access tunnel into the mountain. Through this tunnel, rock blasted from the reservoir was moved out—the equivalent of 69,000 big freight-car loads of it. In addition, more than three miles of tunnels were constructed to provide for the flow of water to the mighty turbines that utilize the power of the stored water.

At one time few power experts would have believed that hydroelectric power could possibly compete in an age of atomic energy. Today it is proving

How Will We Meet the Energy Crisis?

that it has a greater potential than most authorities thought. Tomorrow, it is safe to predict, this potential will be even greater. Perhaps we can even harness the power that is locked in the world's restless oceans.

8
POWER FROM THE SEA

A factory in the sea? The humming machinery in the fish-processing plant testifies that it is indeed that. A farm? It's that too—a farm that raises fish, kept confined by barriers of electric "fence." A floating apartment building? Yes; it has comfortable living quarters for the island's hundred workers.

But it is something even more remarkable. It's a power plant. The block-long row of generators are turned by steam from the sea.

Energy in the Tides

Drawn by the gravitational field of the moon and, to a lesser degree, by the pull of the sun, our ocean waters roll round the world in a giant wave. In mid-

How Will We Meet the Energy Crisis?

ocean this heaping of water is almost unnoticeable, but when tides meet land, they pile up spectacularly. If the tides roar in between the narrowing walls of a bay or up a tidal estuary, they can rise to 30-foot, 40-foot and even 50-foot heights.

Twice every day or, more precisely, twice every 24 hours and 50 minutes (the time it takes the moon to complete one revolution of the earth) the tides rise and recede. Today, bold engineers have taken giant steps forward in their efforts to tame the ocean's tides and use their mighty force to generate electricity.

Men have been using tidal power, in a small way, for 1,000 years or more. European coast dwellers

Can the restless power of the tides be harnessed to provide electricity? Engineers believe that there are many parts of the world where "moon power" can be put to work.

MASS. DEPT. OF COMMERCE

utilized the flowing of the tides to drive millwheels that ground grain. Their engineering achievement was simple, yet it is basically the same as that used in the biggest tidal power projects of today. They placed gates across inlets to small tidal basins and allowed a rising tide to enter. When the tide receded, a pool of trapped water was left behind. Channeled out through a spillway, this water could turn a millwheel.

In the early part of this century, as the need for electricity climbed, many engineers became excited by the possibility of using the tides to produce power. They studied tide charts and maps of coastal areas, looking for locations where large tides could be trapped by dams. Their plan was simple. They would build a dam across a large tidal basin, trapping millions of tons of water when the tide rushed in. Then the water would be let out through channels in the dam, driving giant turbine wheels and generating electricity in great quantities.

Power in Passamaquoddy Bay

In Passamaquoddy Bay, which forms the border between the state of Maine and the province of Nova Scotia, nature has placed a string of islands across the bay mouth. Incoming tides roaring up the larger Bay of Fundy race through narrow channels between these islands with earth-shaking force. The difference between high and low water levels can be more than 40 feet and averages an impressive 26 feet. That rushing water could produce a lot of power!

In 1919 an American engineer, Dexter Cooper, proposed a bold plan for producing power in Passamaquoddy Bay. The islands provided by nature would form much of the dam needed to turn the bay into a closed pool where the tides could be trapped. Cooper was convinced that a tidal power project in Passamaquoddy would give cheap electricity. While a tidal power plant would be expensive to build, it would be cheap to run, requiring no fuel.

Scoffers said Cooper's plans would never work, but he stubbornly persisted in his efforts to get backing for the project. In the early 1930s, President Franklin D. Roosevelt became interested in Cooper's proposal, and pushed the United States Congress to appropriate some money for it. In 1935, work began at Passamaquoddy Bay. Barrackslike buildings went up to house the men who would work on the great power project. A dike was built, beginning the enclosure of the bay. That dike still stands today, strong after decades of buffeting by tides and wind, but no more work has been done in Passamaquoddy. Political objections to the project caused Congress to refuse more money.

The French Dream of Tidal Power

Even before the Passamaquoddy project was born, French engineers were working on tidal power. An early French experiment that worked is almost forgotten now. In the 1920s the French put a small turbine into a dam in the estuary of the river Diouris, on the western end of France, just a few miles north of the well-known port of Brest. This turbine, coupled to a generator, produced a fair amount of current, most of which was used by the government arsenal in Brest. But it was small, and what the engineers really wanted was a big plant that could produce tens or even hundreds of thousands of kilowatts.

Why were the French so interested in tidal power? Because nowhere else in the world are there so many possible locations for tidal power plants jammed together in so few miles of coastline. The western shores of France obstruct the tides moving in from the Atlantic, and large tides occur almost everywhere on the coasts of Brittany and Normandy. At Brest harbor, tidal range is 21 feet. Working around the coast to the north and east, tides get steadily larger. They average 27 feet where the river Diouris flows into the sea. They are more than 37 feet in the estuary of the Rance, and pile up 41 feet in the Bay of Mont-Saint-Michel. All along this coast of high tides there are bays, estuaries and inlets, and at one

FRENCH EMBASSY, PRESS & INFORMATION DIV.

Rance Dam under construction. Giant floating cranes were used to handle heavy components.

E.A.G.: DEPARTMENT PHOTO-INDUSTRIELLE

One of the 24 bulb turbines that turn the tides into power in the Rance Dam. Here the turbine is in place inside the dam. Water flows through the blades in the background, driving the turbine.

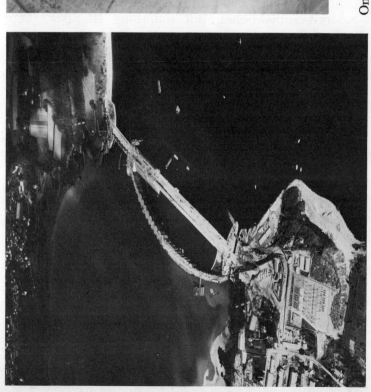

FRENCH EMBASSY, PRESS & INFORMATION DIV.

The Rance Dam as it looked when near completion.

time or another someone has proposed a tidal power project for almost every one of them.

In 1940, Robert Gibrat, Director of Electricity Distribution for France's Ministry of Public Works, read some of the old discusions of tidal power projects. The more he read and thought, the more excited Gibrat became. The son of a Brest physician, he had spent his boyhood on those French coasts where the tides run so high, and his memories translated the dry figures and drawings on paper into vivid pictures of what could be.

Suppose he were going to build a power plant on the French coast. Where would he put it? Gibrat was, of course, impressed by the 41-foot tides at Mont-Saint-Michel, but damming the bay there would require building two dikes, a total of 25 miles in length. No; the river Rance was the place. Tides were a little lower, but there was a spot where a half-mile-long dam could span the estuary. Put a dam across that point, one equipped with gates and turbines. Open the gates when the tide flowed in, and then close them at high water. Later, when the tide outside the dammed basin was low, allow the water in the basin to flow out, driving the turbines and producing electricity.

Gibrat, enthusiastic about his project, knew that it was a job requiring the collaboration of many scientists, and he managed to fire others with some of the enthusiasm he felt about tidal power.

At first, work was discouraging. Calculations seemed to show that tidal power could not produce enough electricity to justify the enormous cost of building a dam and power plant across the Rance. Gibrat, however, was determined.

"The energy is there. All we have to do is find the right way to turn it into electricity," he said.

The main problems stemmed from the tides themselves. First, there was the annoying fact that tides do not operate on a neat 24-hour schedule, as do human beings and their power consumption, but on that 24-hour, 50-minute lunar cycle. A tidal power plant would produce the greatest amount of electricity if it could run with the tides, beginning to generate whenever the difference between water level inside the dam and outside the dam was greatest. But men function on solar time, not lunar time. Some way would have to be found to make tidal power work on a 24-hour schedule.

Then there was the problem of neap tides. Tides, unfortunately, are not all the same height. At full moon and new moon, the sun and moon pull together and large tides occur. But halfway between these large, or spring, tides, solar and lunar gravitation tend to cancel each other's effects, and the neap tides, much lower than others, occur. A power source that produces a lot of power one week and not so much the next has obvious disadvantages. An engineering solution had to be found that would en-

How Will We Meet the Energy Crisis?

able the plant to produce about the same quantity of power every day.

The New Bulb Turbine

Among the many groups Gibrat talked into working on the Rance project was the Société Grenobloise d'Études et d'Applications Hydrauliques (SOGREAH). There a team of engineers, headed by Jacques Duport, began working to find a new kind of turbine.

The classical turbine for handling large quantities of water is the Kaplan-type turbine. Its propellerlike blades are mounted on a vertical shaft. The angle at

The tide came in. By channeling this surging water through turbines, we can turn tidal energy into electrical power.

CANADIAN CONSULATE GENERAL

which the blades "bite" the water can be changed while the wheel is running to adjust to differing amounts of water flow. This turbine works well in a river dam, where the water runs the same way all the time, but Gibrat, Duport and other engineers knew that a single-basin tidal power plant working with water flow in only one direction could not make electricity cheaply. They had to be able to pump in both directions. Trying to solve that complicated hydraulic problem evoked hundreds of ideas and patents, some of which did not work well and some of which were too expensive.

Duport describes SOGREAH's work like this: "Starting about 1943, we focused our attention on a quite new approach—the development of a single piece of machinery that would sit right in the water-flow channel and perform the different pumping and turbining tasks. What emerged after much travail was the 'bulb unit.' It looks like a small submarine and is composed of a metal shell containing an AC generator of special design. The axis of the unit is generally horizontal, and the water circulating around it drives a turbine at one end.

"This hydroelectric device is quite versatile. It can function as a turbine with water flowing from either direction. It is, in fact, the first and only reversible turbine. When necessary, the generator can be used to work as an AC motor so that the bulb unit can function as a pump, also with water flowing in

How Will We Meet the Energy Crisis?

either direction. Finally, the bulb unit can be used as a gate in either flow direction; thus the number of sluice gates in the tidal power dam can be minimized."

In 1959 the French tested the first bulb turbine by installing it in an unused harbor lock at Saint-Malo, a city located where the Rance flows into the English Channel. The turbine performed better than expected. It produced 9,000 kilowatts of power. (The 24 bulb units in the Rance dam each have a 10,000-kilowatt capacity.)

The bulb turbine was a revolutionary breakthrough in turbine design. It is the machine that makes large tidal power projects possible.

The Problems of Damming the Rance

Construction engineer might well have been dismayed when they stood on the banks of the Rance and watched the foaming waters rush in and out with the changing tides. No one had ever dammed a large tidal estuary before, and the problems were formidable. Structures damming a river must withstand only a one-way stress. Here the stresses would come from both directions. The channel was only 2,460 feet wide at the point selected for the dam; at midtide, water races through this narrow channel at the rate of 635,400 cubic feet per second.

To hold back the surging Rance waters, engineers

first built 19 giant caissons, cylinders of prestressed, reinforced concrete, the largest of them 30 feet in diameter and 85 feet long. One end of each caisson was closed. As they were towed out into the river they floated like huge bottles. On the river floor, concrete bases were ready to receive them. The caissons were guided into place by chains. Tons of sand poured into them caused them to sink to the bottom, where they stood upright. The spaces between them were then closed off with sheet-metal piles. With the tides thus held in check, engineers went on with the construction of the final dam, which includes a six-gate sluiceway and a navigational lock as well as the power plant housing the 24 bulb turbines.

The Rance Power Plant Today

Today the great dam across the Rance houses a plant that operates on a varied schedule to extract the greatest possible amount of power from the tides. Usually the sluice gates are opened after low tide. As the tide rises, water rushes in through the gates and fills the basin above the dam. When the filling process is completed, the gates are shut.

Later, the turbine channels are opened and water rushes through them back to the sea. The turbines spin, the generators turn, and power from the tides surges into the French transmission network.

How Will We Meet the Energy Crisis?

Sometimes, in order to provide power for periods of "peak need," the turbine channels must be opened on the rising tide. The reversible turbines produce power with water flow in this direction. At other times, to get enough power out of the low neap tides, the turbines are used as pumps, to raise the water level in the basin before letting it out again to generate electricity. The French discovered that this pumping is very profitable. Pumping water into the basin when water levels in sea and basin are almost the same takes little energy, and many times that expenditure is returned when the heightened basin water drives the turbines.

The Russians Begin at Kislaya Inlet

Interest in tidal power is intense in many countries, but the Soviet Union, hungry for new sources of power, was the first country to follow the French lead. Shortly after construction began on the Rance, Soviet engineers went to work on a pilot project at Kislaya Inlet, on the Kola Peninsula near Finland. This 1,200-kilowatt-capacity plant is only the small beginning of Soviet plans. They hope that tidal power will help them achieve their goal of 500,000- to 600,000-kilowatt capacity by 1980.

Kislaya Inlet presented new problems, ones not solved by the French. Tidal variation in the 100-

foot-wide channel that links the small, deep basin with the sea is only 13 feet. To generate electricity with this lesser difference in water levels, the Russians had to gear their turbines to spin much faster than the ones in the Rance. The turbines, however, are of the same general design. As a matter of fact, the first of the three 400-kilowatt units was built by Neyrpic, the French company which built the turbines for the Rance project. The Russians built the other two themselves, to a slightly modified design.

To dam the inlet, the Russians used prefabricated sections of concrete dam, floated to the site in the same fashion as the caisson posts used by the French.

There are many locations on the northwestern shores of the U.S.S.R. where tides range from 26 to 30 feet in height. After Kislaya, the Russians plan to move on to Lumbov, on the Arctic Ocean at the northeast curve of the Kola coast. There a small dam will wall off a 27-square-mile bay, and the plant to be built will have a capacity of 320 megawatts. The Russians will build a larger plant in the estuaries of the Kuloy and Mezen Rivers, across the White Sea straits from Kola. In the end, the Soviet dream is to build the biggest plant in the Mezen Bay. This system of plants will provide great quantities of cheap power to meet power demands across the entire span of European Russia.

How Will We Meet the Energy Crisis?

Tidal Power for Tomorrow

In a global survey, engineers have pinpointed nearly a hundred possible sites where electricity could be generated by the ocean's moving waters.

One proposed plant, that on the Severn River in England, might be hooked into a power network with the Rance River dam and the talked-of Mont-Saint-Michel installation. Forty-five-foot tides in the Severn offer promise of great quantities of electricity. Since tides on the English coast run at different times than those on the French coast, a unified power network might provide a constant flow of power for England and France. This would help eliminate the one serious disadvantage of tidal power —the periods when a plant is not producing power.

Other sites are found in many countries—including Australia, South Korea and Argentina. With the exception of Passamaquoddy, the continental United States has no possible locations. There are, however, places in Alaska where high tides rush into river estuaries and bays.

On the coast of Argentina are two bays with narrow openings that seem nearly perfect for tidal power. They are Bahía Nueve and the Gulf of San José, on opposite sides and at the base of the Valdés Peninsula on the Atlantic coast. The Gulf of San José could be closed by a dam 4½ miles long. The dam would be hard to build, but total power

CANADIAN CONSULATE GENERAL

The power of the tides is dramatically illustrated by this photo and the next. Here, on the coast near Parrsboro, Nova Scotia, it is low tide.

Aerial view of proposed Passamaquoddy tidal power site. The dams and turbines that might be built here someday would harness the mighty tides that surge up the Bay of Fundy.

U.S. DEPT. OF THE INTERIOR

output might reach as high as 5,500,000,000 kilowatt-hours per year. It would take the burning of more than 2,000,000 tons of coal to produce that much power by conventional means.

Power From the Oceans' Surface Water

As the sun beats down on the oceans of the world, it heats the surface waters. In tropical and subtropical regions, water temperatures climb as high as 80° or 90° F. In the form of this heat, the oceans store vast amounts of solar energy.

The power-producing potential of warm water is so tremendous that it dwarfs tidal power by comparison. After all, there are a limited number of places where a tidal power plant can be built, and only so much electricity can be generated at each of them. Successfully tapping the energy riches locked in warm water could supply power for the entire world.

The first steps have already been taken. At Abidjan, on Africa's Ivory Coast, a small power plant generates electricity through a process that makes use of the temperature difference between surface water (heated to 82° F. by the tropical sun) and bottom water (a constant, chilly 46° F.).

The warm water passes through a boiler, where it is turned into steam, without the application of more heat. How can water boil at a mere 82° F.? The secret is low pressure. That familiar figure, the 212° boiling point of water, is the temperature at which water boils at ordinary sea-level pressure. Lowering the pressure lowers the boiling point of water.

When the warm sea water turns into steam, it expands, rushes through a steam turbine and drives the turbine wheels. After the steam passes the turbine, it enters a condenser. There the cool water pumped up from the ocean chills the steam, turning it back into water. Now, of course, the water occupies less space than it did as steam, and this helps produce the vacuum which makes the water boil in the first place. It also helps to pull the steam through the turbine.

Engineers have discovered that, unfortunately, much of the electrical output of the Abidjan plant is consumed in pumping the cold water from the condenser. But they have proved that power can be generated from temperature differences in the sea, and experience gained at Abidjan will be invaluable in future projects.

Experimenters say that using the sea water itself to drive the turbines is not the most efficient way of exploiting the heat energy locked in the oceans. Other substances, such as propane, are much denser than water at low temperatures, and use of them would permit smaller, less bulky and therefore less costly turbomachinery. Propane has another advantage. It is far less corrosive than the salty waters of

How Will We Meet the Energy Crisis?

the sea. In a propane-using, sea-thermal power plant, the warm surface waters would heat the propane to a boil in a low-pressure chamber, and the cold bottom water would cool it back to liquid.

One of the greatest expenses in building the Abidjan plant was the laying of the cold water pipe far out from shore along the sloping sea-bottom. Engineers looking toward tomorrow have plans for ocean power plants that would be giant floating platforms. From them the cold water intake could drop straight down to levels where the ocean was cool enough for efficient power production. These floating power plants would be anchored to the ocean floor and would send their power to land by submarine cable.

Scientists estimating the potential of sea-thermal power figure that the Caribbean Sea and the Gulf Stream can yield 182,000,000,000,000 kilowatt-hours of electricity. That is some 60 times the estimated power consumption of the United States in 1980. World-wide, the floating power plants of tomorrow could provide two hundred times our estimated global electrical needs in the year 2000.

9
POWER FROM THE EARTH

"Zero minus ten . . . nine . . . eight . . ."

As the voice booms over the loudspeaker, a restless murmur runs through the crowd of scientists, engineers and reporters.

"Seven . . . six . . . five . . ."

The spectators who look out across the rolling hills of upstate New York are tense with excitement as the countdown continues. This is one of the big moments in the history of science. Buried thousands of feet underground is a powerful nuclear explosive that, the waiting scientists believe, will release energy locked in the earth.

"Four . . . three . . . two . . . one."

The ground rocks gently under the feet of the

How Will We Meet the Energy Crisis?

watchers—the only sign that far below a fiery explosion has unleashed a vast new source of power

Power from the earth!

This is a dream that has long excited men who have witnessed the flaming fury of volcanoes and the roaring force of steam geysers. Today it is a dream that is becoming reality, showing great promise of providing power for tomorrow's world—power without pollution.

The earth offers man a deep reservoir of heat energy. As miners long ago discovered, the depths of the earth are warm. In general, temperature rises 1° F. for each 60 feet of depth. Geologists theorize that beneath the thin, cool crust of rock on which we live is a hot layer of molten rock they call the mantle. Temperatures in the mantle are probably about 5,000° F.

Fortunately, we are shielded from the mantle's intense heat by miles of insulating rock. Beneath some of the world's highest mountain ranges, the crust may be as much as 40 miles in thickness. In some places the mantle's heat rises through weaknesses in the earth's crust, and temperatures of more than 1,000° F. can be found within a few thousand feet of the earth's surface. These "hot zones" are usually associated with recent volcanic activity and are often marked by hot springs and geysers, formed by the heating of ground water. By drilling wells to

PACIFIC GAS & ELECTRIC CO.

Young tourists witness the evidence of power in the earth, as engineers prepare to connect eight steam wells at The Geysers to generators.

release steam, modern mining and power engineers are finding ways to use these thinly buried heat reservoirs to produce electricity. The steam can be used to run turbines just as would steam produced by burning fossil fuels or fissioning atoms.

The pioneering project for using thermal energy to produce electricity was carried out in Italy. The countryside around the city of Larderello was dotted with bubbling sulfurous springs and whistling vents of steam rising from cracks in the hard red earth. In the 19th century, some of this steam was used to run machinery in factories set up to extract chemicals from the volcanic waters. At the beginning of the 20th century, engineers began to wonder if they couldn't use the steam to generate electricity.

Once they had the idea, it was not difficult to rig up a system. By 1904 they had built a small power plant, consisting of a ¾-horsepower steam engine which operated a dynamo. This dynamo produced enough electricity to power a few lightbulbs, and it was the beginning of what is now called geothermal power. Soon another engine, this one with a 40-horsepower rating, was producing enough electricity to light the whole town of Larderello. The Larderello steam industry grew swiftly as engineers drilled "steam wells," trying to hit pockets of underground steam more powerful than the vents supplied by nature. Today more than 150 steam wells are helping produce power for nearby cities.

Power From the Earth

In spite of its success, the Italian venture was little more than a curiosity until recently. Other ways of power production were cheaper. Half a century passed before the next geothermal power plants went into operation—one in New Zealand, the other in the United States.

Like the steam jets of Larderello, the site of the United States plant had long been known as a natural wonder. In 1847, a year before the discovery of gold in California, an American surveyor named William Bell Elliot was hunting grizzly bears when he wandered into a valley in Sonoma County, 85 miles north of San Francisco. What he saw in that valley drove all thoughts of the chase from his mind. Plumes of steam waved in the wind as they issued from dozens of vents in the earth. The valley was a veritable inferno. A later writer described it as a "branch of Hades."

"The Geysers," as the valley came to be called, was for a long time a popular sightseeing attraction and resort, famed for its hot mineral baths. In the early 1920s the enterprising proprietors of the resort harnessed the steam to run a generator which provided electricity for lighting the hotel buildings. It was hardly a great success, because the electricity cost more to produce than hydroelectric power.

Thirty years later, when the scramble to find new sources of power had begun, engineers made another effort to tap the thermal resources of The

How Will We Meet the Energy Crisis?

Geysers. The plant they completed in 1958 dwarfed the first one, producing enough power to make the engineering world really consider the possibilities of harnessing the heat locked in the earth.

Today the valley of The Geysers has more visitors than it ever did in the days when it was a resort. They come to see the power plant which, using steam from inside the earth, sends more than 200,000 kilowatts of power surging into the transmission lines that carry electricity to many parts of northern California served by the Pacific Gas & Electric Company.

Other geothermal power plants are now operating in many parts of the world. New Zealand, Mexico,

Steam from the wells is piped to the power plant.

PACIFIC GAS & ELECTRIC CO.

PACIFIC GAS & ELECTRIC CO.

Engineers test a mile-deep steam well for possible use as a source of power at The Geysers.

How Will We Meet the Energy Crisis?

Japan and the Soviet Union are countries with actual plants. Many more are looking into the possibility. Engineering studies, some sponsored by the United Nations, are being made in Chile, Costa Rica, El Salvador, Ethiopia, Guatemala, Hungary, Israel, India, Kenya, Indonesia, Jordan, Nicaragua, the Philippines, Turkey and the United States. All of these countries have promising geological formations which could be readily tapped for their heat resources.

As they begin to look closely at the earth as a source of power, engineers find themselves awed by the vast potential. At a world-wide conference of scientists meeting at Pisa, Italy, more than 200 re-

Japan is among the nations working to develop geothermal power. Here, steam rises from cones in a volcanic region.

UNITED NATIONS

ports were presented describing the results of studies made in various countries. One, made by Soviet experts, indicated that the geothermal potential in that country is greater than all other energy sources put together. Other reports indicated that the geothermal possibilities of the Jordan River in Israel and Jordan could provide electricity for the entire Middle East. (Of course, capitalizing on this possibility would depend on an Israeli-Arab peace.)

One of the greatest of geothermal bonanzas is right here on our own North American continent, in the western part of the United States and northern Mexico. Geologists have determined that a long-past cataclysm in the Pacific Ocean may be used to provide electricity for your home! Fairly recently as geological dates go—perhaps as little as a million years back—a welling up of magma from the earth's molten interior pushed up under the floor of the Pacific Ocean, near the North American continent. This is not unusual; it occurs in other oceans, too. But in them it simply pushes lava up into the mountain ridges under the sea. The enormous blister of congealing magma under the Pacific floor has been slowly pushing its way eastward, *underneath* the North American continent, causing the peninsula of Baja California and most of southern California to be slowly pulled away from the continent.

The effects of this stupendous geological event are not limited to the West Coast. The blister is ris-

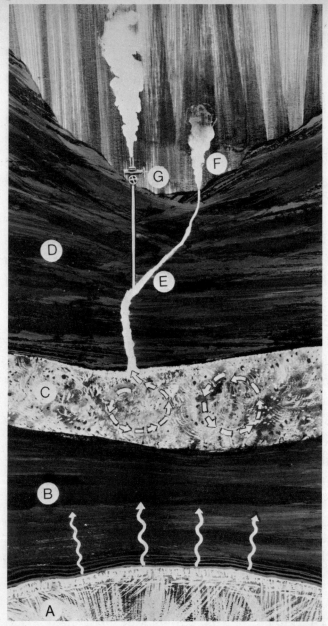

Diagram showing a geothermal field. A. Magma. B. Solid rock conducts heat upward. C. Porous rock contains water that is boiled by heat from below. D. Solid rock prevents steam from escaping. E. Fissure allows steam to escape. F. Geyser. G. Well taps steam fissure.

ing under the whole western part of the continent, perhaps as far east as the Mississippi River. It is the cause of hot springs that occur all over the west. Even where no such hot springs are visible, the earth relatively close to the surface is being heated up by the pressure underneath.

Under this vast area of western North America, geologists picture huge quantities of steam, just waiting to be tapped. One of President Nixon's final acts on 91st Congress legislation was the signing into law of the Geothermal Steam Bill. This opens up some 1,350,000 acres of federal land to exploration, development and use of geothermal steam.

A large area in Mexico's state of Baja California and in California's Imperial Valley is destined to be the first point at which power engineers will attempt to extract the heat made available by this monstrous blister under the earth.

In 1939 a Mexican engineer, Luis F. de Anda, who loved to hike around his country's mountains, got to thinking about the numerous springs he had encountered in his wanderings. As he watched Indians cooking potatoes in these hot springs, he thought of the power plant at Larderello, Italy. Perhaps his own country had a hidden treasure that no one thought about using. De Anda talked it over with a noted geologist, Frederico Mooser, at the University of Mexico. Mooser thought the idea of drilling for steam power was a good one, and eventually the Federal Electricity Commission (CEF) backed the drilling of a steam well near Mexico City. Meanwhile, an engineering report came in from a CEF engineer who had observed the springs around Mexicali in Baja California. There must be a lot of hot water under the surface, he concluded.

That turned out to be one of the understatements of the century. When De Anza and his engineers moved in to really study the region, they emerged with a stunning geological picture of the Salton Trough. They determined that underlying a 1,000-square-mile area around the mouth of the Colorado River and extending under California, is a huge rocky "sponge" reaching down 20,000 feet beneath the earth's surface. The amount of water trapped in it is so great that if it were on the surface it would cover as much as 10,000,000,000 acres to a depth of one foot!

On the U.S. side of the border, researchers are laying plans to drill the wells to tap this immense resource of hot water. At the University of California at Riverside, a team of international scientists, headed by Dr. Robert Rex, is making the most detailed study ever made of the geothermal potential of any area. By the 1980s, some optimistic reports indicate, the huge reservoir of trapped hot water could be providing enough power to meet the growing needs of all of southern California and northern Mexico.

U.S. NATIONAL PARK SERVICE

Steam and hot water burst from Castle Geyser in Yellowstone National Park. Steam power which may be tapped to produce electricity underlies much of the western U.S.

How Will We Meet the Energy Crisis?

Man-Made Boilers Under the Earth

Going beyond their hope that hot water underground can provide pollution-free power for tomorrow's world, scientists have figured out other ways to use the earth's heat, even where there is no large amount of water in the ground. A way of doing it was proposed some time ago by a British scientist, Sir Charles Algernon Parsons. Why not, Sir Charles asked, dig a shaft in the earth and hollow out a large cavity at the bottom which could serve as a great underground boiler? He reasoned that if such a cavity could be made, you could then insert two pipes down the shaft. Down one you could pour water which, in its fall, would drive turbines. When the water reached bottom, the heat in the "boiler cavity" would flash it into steam, which would rise up the second pipe, driving turbines on its way. Sir Charles thought that such a system could give man an unlimited supply of cheap power.

The idea was very sound, except for one insurmountable obstacle. In the early part of this century, when he proposed it, there was just no way men could carry out that kind of deep mining in hot rock. There was speculative talk of automatic machinery that would go down and hollow out the cavity, but no one knew how to build it.

Of course, that was before the days of applied atomic energy. Today, scientists are sure that nuclear explosives could be used to blast a cavity of enormous size—and, at the same time, create great additional heat to add to that of the earth. A complete plan has been worked out by Dr. George C. Kennedy, Professor of Geochemistry at the University of California.

This is the way Kennedy's heat mining plan would work: Drillers would sink a shaft three feet wide down to a depth of 10,000 feet, where rock temperatures are around 1,000° F. It would be an ambitious project, but it would utilize skills and equipment already developed for oil-well drilling. Then AEC technicians would implant a nuclear "bomb" at the bottom of the shaft, one of five-megaton size (having explosive force equal to that of *5,000,000 tons of TNT*).

AEC experiments in the Nevada Test Range, 85 miles north of Las Vegas, show what an underground blast of that force would do. The initial explosion would create a giant cavity in the rock, perhaps 1,000 feet in diameter. Afterwards, rock would fall in from the cavity roof and a chimney of shattered rock would reach upwards from the cavity, extending to within less than a mile of the earth's surface.

This chimney of rock would conduct heat upwards from the 10,000-foot detonation level. Sinking a second shaft to reach that chimney would be much easier than drilling the first. Extracting the heat energy as steam would be done with a variation of Sir Charles's simple two-pipe plan. One pipe would introduce water into the hot rock, and the other would recover the steam.

Professor Kennedy offers some figures indicating how much power this kind of nuclear blasting could produce: "There will be approximately 4,000,000,000 cubic feet of rubble in the chimney. The mean temperature of the rubble will be approximately 350° C. We could obtain 4,500,000 calories of heat from each cubic foot of rock. Altogether, there will be available some 18,000,000,000,000,000 calories of heat." Only one-sixth of that energy bonanza would come from the bomb—the other five-sixths would come from hot rock inside the earth.

Setting off one five-megaton blast in the right kind of rock would provide energy to produce 100,000,000,000 pounds of steam. Purer, hotter and more valuable that any steam from natural vents, this steam would be worth a minimum of $10,000,000.

Estimates are that drilling the shaft would cost $4,000,000, and that the AEC would charge $1,000,000 for the bomb. Thus, there would be a $5,000,000 profit. The steam from the first explosion would run a 50,000-kilowatt plant for ten years, and still there would be heat left in the chimney.

Even that small remnant of heat can be extracted by blasting a second chimney right beside the first. The second chimney would contain as much heat energy from the bomb and from the earth as did the

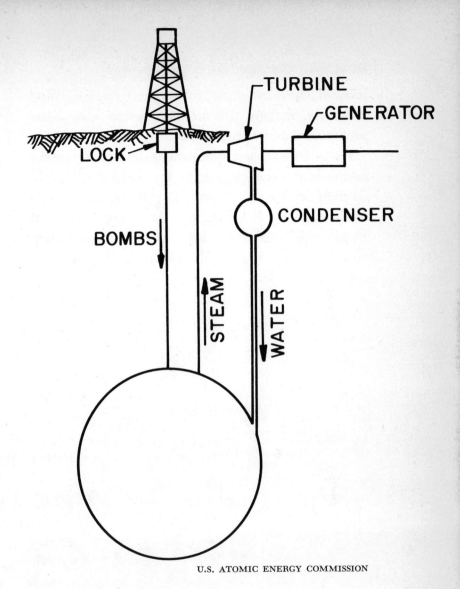

How "atom bombs" might be used to produce power anywhere in the world. Nuclear explosives lowered through drill-holes would be set off, forming a very hot underground cavity. Water pumped in through another hole would turn to steam. The steam would rise through a third hole, driving a turbine.

How Will We Meet the Energy Crisis?

first, and heat lingering in the first chimney would leak over, making this second nuclear blast more profitable than its predecessor.

Even if we do not use nuclear devices to tap the earth's heat riches, engineers say that by 1980 geothermal power will be providing 10 per cent of the world's power. If we *do* use the mighty forces of "bombs for peace," the heat under our feet could give us nearly unlimited quantities of power for countless centuries.

10
POWER FROM THE SUN

"*Ladies and gentlemen, we are interrupting your regular entertainment to let our cameras bring you a close-up view of the landscape below. We are now approaching the Sahara Solar Development Region.*"

At the pilot's words, the passengers in the supersonic jet turn their attention to the full-color 3-D screens. At first a scene of desert desolation appears —nothing but tawny sand. Then a green strip comes into view at the top, grows, and soon fills the whole screen, as the wide-angle telescopic lenses of the cameras reveal thousands of square miles of verdant cropland.

The travelers stare with interest at this area they have heard and read so much about—the fabled million-acre man-made oasis. They know that the

How Will We Meet the Energy Crisis?

fierce sun which once seered this land and made it inhospitable for life has now been put to work by science to turn it into one of the garden spots of the world. Electricity produced by solar energy powers the towns and farms and mighty pumps which draw up abundant water from deep in the earth.

This picture of a future development may not come true until the end of the century, but it seems certain that man will turn to the sun for electricity well before that time. The blazing hydrogen furnace of the sun is the most abundant potential source of power for our planet. It can provide far more energy than we could ever hope to obtain from all the fossil fuels and rushing rivers in the world. The possibilities of solar power, in the opinion of many researchers, dwarf even those of atomic energy derived from fusion.

The amount of energy received from the sun is awesome. Every *20 minutes* enough solar energy strikes the United States to meet all our power requirements for an *entire year*. It has been calculated that in a single year the small part of the earth represented by the Sahara Desert receives a thousand times the amount of energy that would be released from burning all the known coal reserves on our planet.

The problem is to find a way to capture this energy and turn it into electrical power. Already in

UNITED NATIONS

A simple use of solar energy. Sunlight, reflecting off the polished metal surface of this solar cooker developed in India, is concentrated on the rack, where food is cooked.

many parts of the world the sun's heat is being put to work directly. Houses as far north as Massachusetts are successfully heated by systems using chemicals which absorb solar heat and store it for circulation during periods when the sun is not shining. In sunny parts of the world, roof tanks are used to heat domestic water. In many undeveloped countries lacking both electricity and fuel, cleverly engineered solar cookers give people the equivalent of a modern electric or gas range.

The most spectacular devices using heat energy directly are solar furnaces. They provide researchers with a source of heat far higher than that created by burning fossil fuels. Today's experimenters with solar furnaces have a long history to draw on. The first successful solar furnace was built by the great French chemist Antoine Lavoisier in 1774. In it he used a giant double concave glass lens. More than four feet in diameter, the lens focused sunlight to a very small point. Lavoisier filled the lens with alcohol, which further increased its power. We do not know the maximum temperature Lavoisier achieved with his lens systems, but we do know that he succeeded in melting platinum, which melts at 3,190° F. Lavoisier used this great heat to study substances and chemical reactions at high temperatures.

The inventor of modern solar furnaces was the German scientist Rudolf Straubel. In 1921 he completed work on what he called a "melting oven," to

Power From the Sun

be used in studying metals at high temperature. This melting oven concentrated sunlight by a combined system of mirrors and lenses. When the air was clear and the sun bright, Straubel could subject materials in the oven to temperatures as high as 7,000° F. This temperature is more than sufficient to melt all known solid materials and is, in fact, 70 per cent as hot as the surface of the sun.

The largest solar furnace in the United States is one located at Natick, Massachusetts. It is used by the U.S. Army to test the reactions of materials intended to protect soldiers from the fierce heat of nuclear explosions. While details of the Natick experiment are secret, it is believed that the reflecting mir-

The world's largest solar furnace, in the French Pyrenees mountains, is made up of 9,000 mirrors, which produce intense heat.

ARTHUR D. LITTLE, INC.

ror, 28 feet in diameter, will focus sunlight to a point that may reach 10,000° F.

The world's largest solar furnace is the one designed and built by the French scientist Dr. Felix Trombe, who has been experimenting with solar furnaces ever since World War II. This furnace, located high in the Pyrenees, has a sunlight concentrator made up of 9,000 mirrors.

These furnaces are useful in the study of metals, ceramics and plasmas (high-temperature gases). However, the possibility that someday this heat may be used for producing electricity is always in the minds of researchers using them. The means of doing it has already been developed, in the form of two quite different kinds of devices—solar motors and solar cells.

Solar Motors

Like the furnaces, the solar motor has a long history. The idea was pioneered by a Swedish-born American inventor, John Ericsson, the engineer who designed one of the world's first armored ships—the famous *Monitor*, which fought for the North during the Civil War. The versatile Ericsson was also a pioneer in the field of solar energy. Although he designed several different solar motors that actually worked, none of them ever found practical application. They just didn't produce enough dependable power. Ericsson was far ahead of his time. There was still plenty of coal around to operate steam engines, so nobody could see any point to using the power of the sun.

Later, another American, Frank Shuman, devised several kinds of solar motors. In Shuman's first experiment, completed in 1901, sunlight was concentrated onto a water boiler by a reflector about 30 feet in diameter. This reflector contained more than 1,700 flat mirrors, each positioned carefully to reflect sunlight onto the boiler. This arrangement heated the water, which was then turned to steam.

After many other experiments, Shuman tackled the problem of using sunlight to pump irrigation water in Egypt. When this project was finished, the steam-producing system contained 572 boilers, covering more than 40,000 square feet of ground. Each boiler was heated by its own mirror, which could be turned to follow the sun as it moved across the sky. The steam pressure built up by this complex array of boilers, mirrors and piping was used to drive the irrigation pump. The pump produced an average of 50 horsepower over a ten-hour working day. Construction costs for this water-pumping experiment were high, but operating costs were low, since no fuel was consumed.

The search for cheaper, more efficient solar motors has challenged many researchers all over the world. One Austrian experiment used liquid mercury as

How Will We Meet the Energy Crisis?

the fluid in the boiler. Mercury boils and turns to vapor only when heated to a sizzling 622° F., but sunlight reflected off mirrors achieved this temperature easily.

Because mercury has such a high boiling point, it can produce power not once but twice, as it is cycled through the power-producing system. This is an example of what engineers call a "binary-fluid cycle." In the Austrian experiment, mercury heated beyond 700° F. passed through a turbine. After leaving the turbine, the mercury still retained enough heat to flash water into steam. The water was circulated through the walls of the condensing chamber. The steam produced here drove a steam turbine.

Engineers use a converted searchlight to focus solar energy on a test stand where they are studying the effects of high temperatures on materials.

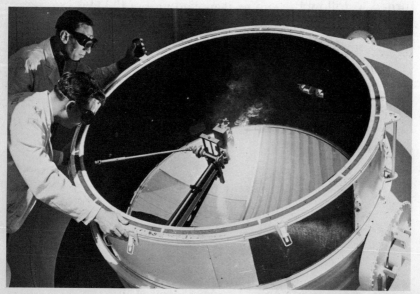

SANDIA LABORATORIES

Other experimental solar engines simply use air instead of a liquid. In them, the sun heats the air, which expands and pushes a piston or spins a turbine. One new engine built by University of Wisconsin researchers is made of transparent quartz instead of metal, so that the sun shines right into the cylinder.

Many technical problems stand in the way of using such engines to produce electricity, but the potential is there. Perhaps they will one day turn giant versions of the small turbines on which they have already demonstrated their capability. Engineers can visualize a system of mirrors, spread over many square miles of sunny areas like the Sahara, heating liquids or air and producing electricity to power pumps and light small cities.

On the basis of present knowledge, it seems sure that the first cost of such equipment would make the electricity produced much more expensive than other means. But in the future, in a world faced with dwindling sources of fossil fuel energy, it may prove worth while to make the capital investment in such power plants.

Solar Cells to Capture the Sun's Energy

Hold a tiny square of silicon, a glasslike substance, on the palm of your hand, and you'll find it hard to convince yourself that you are looking at a potent device that may be the real key to turning the sun's energy into electricity. This little silicon wafer is a remarkable kind of thermoelectric cell—one that can "make" electricity from sunlight.

Scientists in many countries have hit upon various ways of realizing the dream of direct conversion of sunlight to electricity. More than a century ago, Antoine Bequerel, the great French physicist, discovered that when two electrodes are submerged in an electrolytic solution, current will flow when light is allowed to strike one of the electrodes. Since only 1 percent of the light energy is turned into electricity, however, this device, called the photogalvanic cell, didn't get very far.

The photovoltaic cell was much more promising. In 1873 an American telegraph engineer, Joseph May, discovered that under certain conditions, when light fell on a substance called selenium, an electric current was produced. The amount of current is not great, but cells made of selenium found many uses as the familiar "electric eye."

Selenium is one of a class of substances known as semiconductors. They do not carry as much electricity as copper does, but they carry more than nonconductors, such as glass. Research in semiconductors led to the invention of the marvelous Bell Solar Cell, for the scientists who developed it based their device on properties of another semiconductor, silicon.

The Bell experimenters knew that when light

strikes a silicon crystal it "ruptures" the bonds that hold some of the electrons in the crystal to their atoms. (The electron, as you know, is a negatively charged particle.) The bold thought struck them that perhaps they could find a way to gather these loose electrons on one side of the crystal, giving it a negative charge. The side with the diminished number of electrons would then have a positive charge. If a circuit were constructed between the two differently charged sides of the crystal, useful electric current would flow along it, as the negatively charged electrons were attracted to the positively charged atoms across the crystals. An electrical field would separate the positive and negative charges.

After much experimenting, the researchers hit upon an ingenious way to build an electrical field right into the crystal. They made single crystals of silicon which were rich in positive charges on one side and rich in negative charges on the other. They did this by "doping" the crystals with arsenic—one part of arsenic to a million parts of silicon. This created what is called an n-type silicon crystal, one having an excess of negative charges.

Now this n-type crystal was cut into thin wafers with a diamond-cutting wheel. Finally, these wafers were placed in an electric furnace and heated to more than 1,000° C. While being subjected to this heat, one side of each wafer was "doped" with boron. Boron, when it penetrates a silicon crystal, creates a p-type substance, having an excess of positive charges. As researchers had hoped, boron atoms diffused into the crystal.

The final product of this complex process was a thin wafer of silicon that had two layers with different electrical characteristics. This wafer produced its own electrical field, which kept negative charges on the "n" side and positive charges on the "p" side.

The three scientists responsible for the series of breakthroughs that developed these remarkable wafers of silicon were Daryl M. Chapin, Calvin S. Fuller and Gerald L. Pearson. In 1954, when they announced their achievement to the world, they were able to report that the silicon wafers turned sunlight into electricity far more efficiently than any previous method. The best of their original wafers turned 6 per cent of the sunlight striking them into electrical current. They had good reason to hope that even this impressive degree of efficiency could be improved.

The wafers perform their remarkable energy-converting job in this fashion: As photons (individual particles of light energy) strike the silicon, their energy is absorbed by the creation of what scientists call "electron-hole pairs." That is, the light frees an electron to move about in the crystal, and it also creates a "hole," which is merely the positive charge that balances the freed electron. Because of the electrical field built into the wafers, the negative always be above the same spot on earth. Engineers foresee a ring of such satellites, each producing six times as much power as mighty Hoover Dam.

Building the satellites, the receiving stations, the space stations and the shuttle craft to service them is a tremendous and challenging task, but no matter how costly, it would seem to be worthwhile. As a pollution-free source of energy the sun's radiation may offer us the best way to get the electricity we need to fulfill our dreams of a bright new electrified world.

EPILOGUE

"Two powerful forces now at work in American society are headed for a collision that could do damage to both. The first force is the nation's seemingly insatiable appetite for energy to run its factories, commercial establishments, transportation systems, air conditioners, electric toothbrushes and the whole gamut of labor-saving gadgetry and 'modern' conveniences that the American consumer now regards as his birthright. The second force . . . is the environmental movement which seeks to save mankind from smothering in the waste products that result from the generation of energy and from other activities of an industrial civilization. The two forces are not necessarily irreconcilable. . . . The resolution of that conflict will determine whether the nation goes

Epilogue

through a severe energy crisis, a worsening environmental crisis, or both."

This statement in *Science,* the journal of the American Association for the Advancement of Science, summarizes the overwhelming problem that confronts us when we think about power for tomorrow's world. The conflict between these "two powerful forces" brings up scores of questions that must be answered.

Are the natural resources we now use to create power going to hold out if we keep increasing our use of electricity?

Will remaking the map with vast power projects, such as NAWAPA, alter the ecology in ways disastrous to our continent?

Can we really feel safe living in close proximity to atomic power plants?

In using strip mining to get the coal to power huge new thermal power plants, are we going to continue to scar the landscape and ruin vast acreages? Or are we willing to pay for the restorative efforts?

Will we spend the money necessary for research that can develop pollution-free sources of power in the future? Making a determined effort to tap solar power, for example, would cost at least $10,000,000,-000 dollars—perhaps two or three times that amount. Yet in 1970 the total amount spent on research into new power sources was only $350,000,000—only about one-thirtieth of the minimum that would be needed to develop power from the sun.

Or should we do what some ecologists have recommended—give up the American assumption that there's no limit to how much electricity we can produce, and cut down on our use of it?

The making of power is already far and away the largest industry in the United States, two-thirds as large again as the next biggest industry—petroleum refining—and almost twice as large as the giant communications and railroad industries. Conservative estimates are that in 30 years the industry will be seven to eight times as large as it is today! Who should control such an industry—government or private enterprise?

The problems represented by these and many other questions aren't just going to go away. They must be solved, and every citizen has a stake in finding the solutions. No longer can we take electricity for granted, thinking that it will be provided, in one way or another, by distant power engineers, scientists, utility company executives and government bureaucrats. Truly, power is everybody's concern.

It is *your* concern, for you will be living in a world made worse—or better—by the ways we produce and use electricity.

charges migrate to the n-layer, and the "holes" migrate to the p-layer. This creates a difference in potential between the opposite sides of the crystal. Current will flow through a circuit bridging the sides as long as light continues to strike the crystal.

Solar cells have been greatly improved in the years since their invention. Efficiencies as high as 15 per cent are reported. But they are still limited in their usefulness—because of high cost. One of the reasons for their cost is that the silicon used in a wafer must be very pure. Thus, even though silicon (as sand) is one of the cheapest minerals available to man, the silicon in a solar cell is one of the most expensive. Silicon for building a wafer must contain impurities in concentrations no greater than one part in 10,000,000,000!

In spite of their cost, they've proved their usefulness in space, for solar cells have gone aloft in scores of satellites, including the *Telstars*, the satellites that created a new age in communications. The electrical power produced by a satellite studded with thousands of solar cells is capable of keeping apparatus aboard operating as efficiently as if it were plugged into an electrical outlet on earth.

As they witness the performance of these remarkable sources of power, scientists gain confidence that someday they can be put to work to help supply the world's electricity needs. One proposal calls for banks of solar cells catching the sunlight that falls on

BELL LABORATORIES

These solar cells, being assembled for use in a communications satellite, can turn the energy of sunlight into electricity.

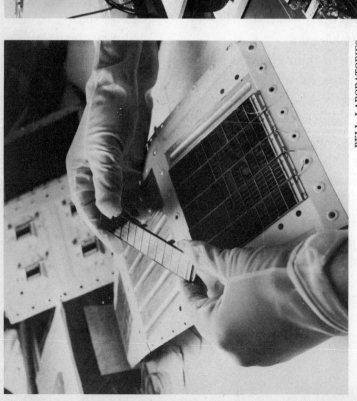

BELL LABORATORIES

A technician tests a solar cell.

large areas of desert. If a cheap way to make the cells can be found, hundreds of thousands of them could be put into such an installation. Any one cell wouldn't produce much power, but a square-mile bank of cells might well supply electricity for a large city.

Engineers have come up with what is literally the farthest-out plan for using solar cells as a means of tapping solar energy. It calls for putting solar power plants in space, where the sun shines all the time. At first glance, the scheme, proposed by Peter E. Glaser, head of engineering sciences at the research firm of Arthur D. Little, Inc., may seem fantastic. However, Glaser and many other authorities are sure that it could be made a practical reality by the end of the century—or even much sooner if we were to start work on it soon.

"It is," says Glaser, "neither more expensive nor difficult than putting men on the moon."

To carry out this great enterprise the engineers would put a giant disk-shaped satellite into orbit. Its 25 square miles of surface would be covered with solar cells. This huge satellite would be assembled in space, by workers using a space station as a base. The current produced by the solar cells would be carried by cable to another satellite nearby, where equipment on board would change the electricity to microwaves.

The microwaves, beamed to earth, would be

How Will We Meet the Energy Crisis?

picked up by receivers consisting of a network of wires. To supply an area such as New York City with power would require such a receiver to be spread out over 36 square miles. Converters would turn the microwaves back into electricity, which would be fed into the regular transmission network. One satellite and one such receiving station would produce 10,000 megawatts of power—one-third more than New York City's entire requirements for 1971.

Any number of power plants could be placed in space. Those orbiting above the equator, 22,300 miles out, would traverse their orbits in exactly the time it takes the earth to turn, so that they would

A visionary scheme? Many scientists don't think so. They are convinced that arrays of solar cells in space can collect solar energy and beam it back to earth to meet the power needs of tomorrow.

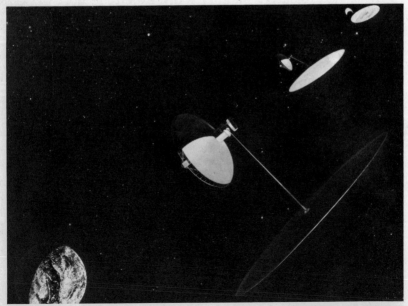

ARTHUR D. LITTLE, INC.

SUGGESTED FURTHER READINGS

Cullen, Allen H. *Rivers in Harness*. Chilton Book Co., Philadelphia, Pa.

Curtis, Richard, and Hogan, Elizabeth. *Perils of the Peaceful Atom*. Doubleday and Co. Inc., New York, N.Y.

Daniels, Farrington. *Direct Use of the Sun's Energy*. Yale University Press, New Haven, Conn.

Halacy, D. S. *Coming Age of Solar Energy*. Harper and Row, New York, N.Y.

Hammond, Rolt. *Power Stations Work Like This*. Roy Pubs., Inc., New York, N.Y.

Landsberg, Hans, and Schurr, Sam H. *Energy in the United States: Sources, Uses and Policy Issues*. Random House, New York, N.Y.

Leach, Gerald. *New Sources of Energy*. Roy Pubs., Inc., New York, N.Y.

SUGGESTED FURTHER READING

McCraig, Robert. *Electric Power in America.* G. P. Putnam's Sons, New York, N.Y.

Millard, Reed. *Clean Air–Clean Water for Tomorrow's World.* Julian Messner, New York, N.Y.

Sporn, Philip. *Research in Electric Power.* Pergamon Publishing Co., Elmsford, N.Y.

SOURCES OF INFORMATION ABOUT POWER

Associations

American Gas Association
605 Third Avenue
New York, N.Y. 10016
American Petroleum Institute
1271 Avenue of the Americas
New York, N.Y. 10020
Atomic Industrial Forum
850 Third Avenue
New York, N.Y. 10022
Edison Electric Institute
750 Third Avenue
New York, N.Y. 10017
Independent Natural Gas Association of America
918 16th Street N.W., Suite 501
Washington, D.C. 20006

Sources of Information About Power

National Coal Association
1130 17th Street N.W.
Washington, D.C. 20036

National Rural Electric Cooperative Association
2000 Florida Avenue N.W.
Washington, D.C. 20009

Solar Energy Society
Arizona State University
Tempe, Ariz. 85281

Government Agencies

Bureau of Mines
U.S. Dept. of Interior
Washington, D.C. 20240

Bureau of Reclamation
U.S. Dept of Interior
Washington, D.C. 20240

National Aeronautics and Space Administration
Washington, D.C. 20546

U.S. Atomic Energy Commission
Washington, D.C. 20545

INDEX

Bacon, Francis, 102
Baker, Donald McCord, 113-114
Bequerel, Antoine, 172
binary-fluid cycle, 170
blackout, 9
brownout, 121

Cerium, 88
Chapin, Daryl M., 174
Cooper, Dexter, 128

Davy, Sir Humphrey, 102
de Anda, Luis F., 155-156
Dupont, Jacques, 134-135

Edison effect, 99
Edison, Thomas, 10
Eisenhower, Dwight D., 87
Elliot, William Bell, 149
Ericsson, John, 168-169

Faraday, Michael, 20
Fermi, Enrico, 37, 39-40, 45
fertile materials, 48

fissionable material, 48
Freeman, S. David, 14
fuel cell, 102
Fuller, Calvin S., 174

Gibrat, Robert, 132-133, 134, 135
Glaser, Peter E., 177

isotopes, 44

Joffe, Abram, 98

Kamerlingh Onnes, Heike, 72-73
Kennedy, Dr. George C., 159-160

Lapp, Dr. Ralph E., 62
Lavoisier, Antoine, 166

Manhattan Project, 41
Mathias, Bernd Teo, 75
May, Joseph, 172
McLean, John, 108

Index

MHD (magnetohydrodynamics), 20-23
Meitner, Lise, 37
millirem, 56-57

NAWAPA (North American Water and Power Alliance), 112-113
Netschert, Dr. Bruce C., 77
niobium, 75
nitrogen oxides, 19
Nixon, Richard M., 155

Parsons, Sir Charles Algernon, 158, 160
particulates, 19
Pearson, Gerald L., 174
photogalvanic cell, 172
photons, 174
plasma, 22, 168
Plutonium, 88, 89, 91
power sources
    atomic energy as, 36-54, 55-67; atomic power plants, factor of safety considered, 55-67; atomic reactors, portable, as, 84-87; breeder reactors as, 46-51; bulb turbines as, 134-136; coal as, 17-26; the earth as, 145-162; atomically-produced electricity as, 41-46; the energy crisis involving, 9-15; the French tidal power experiment and, 128-133, 134; fuels providing, 16-35; fusion, potential as, 51-54; natural gas and, 30-35; geothermal potential for, 145-162; home power plants as, 100-104; hydroelectric power and, 105-124; increased power of nearby hydroelectric plants, 119; long distance transmission of hydroelectric power and, 106-118; Kislaya Inlet and, 138-139; man-made boilers as, 158-162; miniature power plants as, 82-104; nuclear reactors as, 37-41; ocean surface waters, potential as, 142-144; oil and, 26-30; non-polluting fuels and, 16-35; Passamaquoddy Bay and, 127-130; the Rance and, 136-138; the seas as, 125-144; SNAP and, 87-93; solar cells and, 171-179; solar energy, potential for, 163-179; solar furnaces as, 166-168; solar motors providing, 168-171; super-cold-electricity carrier as, 72-78; thermoelectricity as, 93-100; tidal energy from the sea and, 125-126, 140-142; today's demands upon, 9-15; transmission of power, new methods and, 68-81; microwave transmission and, 78; towers for transmission, 69-72;
Project Gasbuggy, 32
Project Rulison, 33
Promethium, 88

Rex, Dr. Robert, 156
Roosevelt, Franklin D., 128

Index

Seaborg, Dr. Glenn, 50
Seebeck effect, 96
Seebeck, Thomas Johann, 93, 96
selenium, 172
Shuman, Frank, 169
SNAP (Systems for Nuclear Auxiliary Power), 87-93
solar furnaces, 166-168
Spilhaus, Dr. Athelstan, 11, 55
Sternglass, Dr. Ernest J., 59

Strontium, 89
Straubel, Rudolf, 166-167
sulfur oxides, 19

Teller, Dr. Edward, 35
Tesla, Nikola, 78
thermoelectric cell, 172
thermoelectric effect, 96
Trombe, Dr. Felix, 168

Washington, George, 30

ABOUT THE AUTHOR

REED MILLARD is an expert in the field of inventions, whose writing ranges from technical reports to popular magazine articles. His deepest interest is environmental science. He believes, as he has indicated in this book, that science can provide solutions to the vast problems that confront us today if we can marshal the needed social energy.

SCIENCE BOOK ASSOCIATES is an organization of writers and technical people active in many areas of science and technology. Their work in the preparation of audio-visual materials and training manuals to be used in science-oriented industry gives the editors an inside look at the developments that will affect tomorrow's world.